XTL

SIMON GOODWIN

XTL

Extraterrestrial life and how to find it

WITH JOHN GRIBBIN

A Seven Dials Paperback

First published in the United Kingdom in 2001 by Cassell & Co

This paperback edition first published in 2002 by Weidenfeld & Nicolson

A CIP catalogue record for this book is available from the British Library

ISBN 1-84188-193-7

Editors: Tim Whiting and Peter Adams
Design: Ken Wilson and Mark Vernon-Jones

Printed and bound in Canale, Italy

Seven Dials
Weidenfeld & Nicolson
Wellington House
125 Strand
London WC2R 0BB

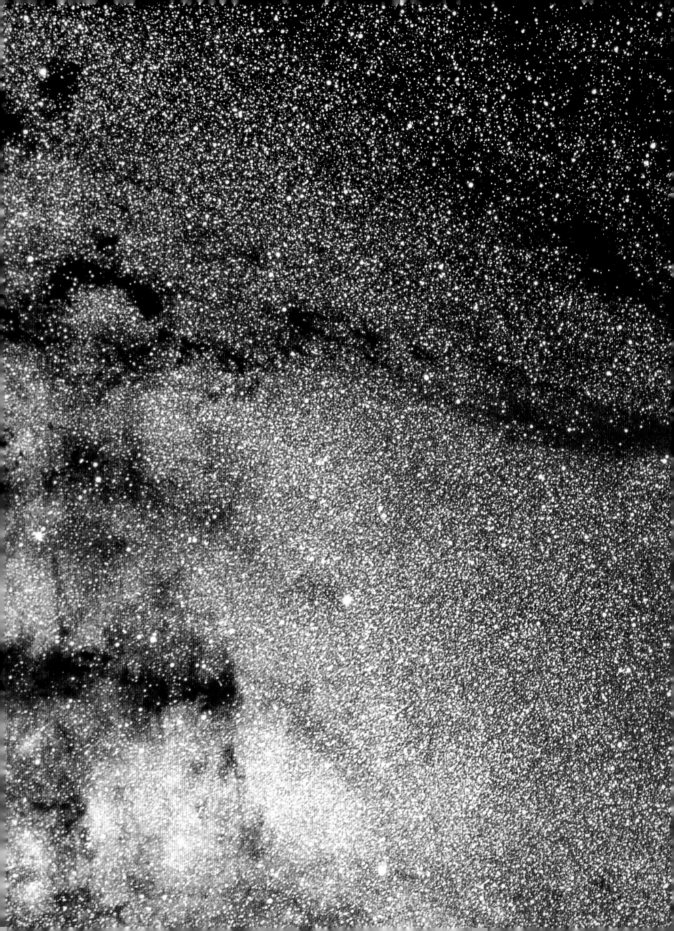

Finding planets

In January 2001, several reports in the media caused a flurry of excitement with the story that two planets similar to the Earth had been found in the astronomical system CM Draconis, which lies 57 light years away from us. Although these accounts were somewhat exaggerated, they were based on real astronomical research reported to a meeting of the American Astronomical Society. The astronomers themselves claimed only that their observations were consistent with the possibility that there might be two planets each about two and a half times the size of the Earth orbiting around the pair of small stars which make up the CM Draconis system. It remains possible that improved observations will show that the effects discovered could be explained in another way; but this work highlights the fact that astronomers at the beginning of the third millennium are on the edge of being able to detect other worlds like our own. It is the story of how astronomers have achieved this ability, and how they know where to look for other homes for life in the Universe, that we tell in this book. Although the evidence for Earth-sized planets remains inconclusive as yet, more than 50 giant planets resembling Jupiter have definitely been identified orbiting other stars. Already one star in 20 close to the Sun is known to have at least one planet and more are being found on an almost weekly basis.

CM Draconis is a binary system; two small, dim stars at the heart of the system swing around each other like a pair of ice skaters holding hands and spinning round and round. If there are Earth-like planets there, such a binary system would make life on them much more interesting than life on Earth. The planets all orbit outside this binary system, and the evidence which supports the idea that there are two planets comparable in size to the Earth comes from a faint flickering in the light from the central system, which could be explained by the repeated passage of the pair of planets in front of the stars, causing tiny eclipses. The exact size of the candidate planets cannot yet be determined from these measurements, but is certainly less than 16,800 kilometres in diameter. This compares with a diameter of 12,800 kilometres for the Earth, and 144,000 kilometres for Jupiter.

All of these observations have been made using telescopes on the surface of the Earth. It is pushing ground-based technology to the limit to identify Earth-sized planets from the ground. But satellites planned for launch over the next few years should be able to test the claims made by the ground-based astronomers, and go on to search for other Earth-sized planets orbiting other stars. The next step, as we describe in detail in this book, will be to test whether the atmospheres of these planets contain oxygen and are likely to be comfortable homes for life. It is entirely possible that evidence for the existence of extra-terrestrial life (XTL) will be available within the next 10 or 20 years. If such evidence is found, it will arguably be the most profound discovery in the entire history of science – indeed, during the entire history of human civilisation. Our hope is to provide you with the background information to enable you to participate fully in the excitement of the search for XTL over the next few years.

Simon Goodwin and John Gribbin

May 2001

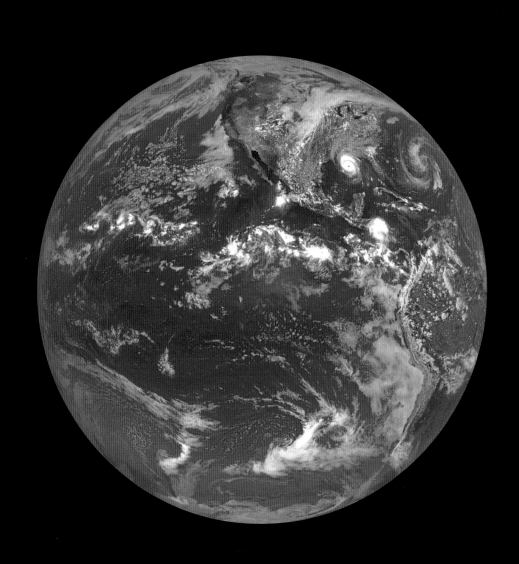

1

our home in space

1.1 The Solar System

The world that we know most about is the one that we live on – the Earth. Since we know that life can exist on a planet like the Earth, it makes sense to start the search for life beyond the Solar System by looking for planets similar to the Earth. As we shall see, there is a real possibility that such planets could be detected within the next ten to twenty years. Even within our Solar System, however, there are a variety of planets, which indicate the range of possibilities that might exist elsewhere in the Universe. By studying these known worlds, astronomers hope to understand just what it is that makes the Earth special. Therefore when they identify other planetary systems they will be able to tell, even across the vast gulf of interstellar space, which of the planets in those systems are possible homes for life and even which currently sustain life.

The most important feature of the Earth, as far as life is concerned, is the presence of liquid water. All living things make use of water, and it is widely thought that life on Earth first started in water, either in the oceans or in what Charles Darwin described as 'some warm little pond'. Just how life started on Earth is an intriguing puzzle in its own right, but one that we do not need to worry about here. The example of our own planet shows that life can exist where there is liquid water, so the first requirement in our search for life elsewhere in the Universe is to find planets with liquid water.

The second requirement, for animals like ourselves, is oxygen, although this is not essential for all forms of life (**see box page 13**).

Furthermore, life also needs a supply of energy. In the case of life on the surface of the Earth, this energy comes from the Sun. Plants use solar energy directly, in photosynthesis, and animals use solar energy indirectly by eating plants (or by eating animals that have themselves eaten plants). But there are forms of life, even here on Earth, which do not use solar energy. Complex ecosystems exist on the seabed, for example, in eternal darkness, making use of energy pouring out from volcanic vents. And microbial life forms have been found living inside of rocks deep below the surface of the Earth, feeding off the outflow of energy from the hot interior of our planet. Light from the parent star is obviously a good

Footnote

Throughout this book when we refer to 'life' we mean 'life as we know it'. Many people have speculated on the possibility of other kinds of life existing elsewhere in the Universe, and we are inclined to think that there probably are forms of life unlike anything we know about on Earth. But until we find them, it seems sensible to stick to what we know for certain.

Oxygen in the air provides animals with the ability to release energy in their bodies by combining material from their food with oxygen – a slow form of burning.

It is this energy which makes it possible for us to move about, keeps our bodies warm, and (many biologists argue) provides the energy needed to develop a large brain. The amount of electricity used to run your brain is tiny; but the amount of energy that goes into building the brain, and keeping it supplied with oxygen-carrying blood, is a major drain on the body's resources.

Oxygen is very important for us, but an atmosphere containing free oxygen is not necessarily essential for life. Indeed, geological evidence shows that there was no free oxygen in the Earth's atmosphere until about two billion years ago, even though there is clear fossil evidence for simple forms of life on Earth in rocks approximately four billion years old.

The oxygen in the atmosphere is actually produced by plants and plant-like bacteria, which use the energy of sunlight in photosynthesis to break down carbon dioxide into their component parts, keeping the carbon for use in its internal chemistry and discarding the oxygen.

Although life can exist without free oxygen, complex active life forms like ourselves need oxygen.

Even better, since free oxygen reacts very rapidly with anything that is combustible (such as carbon), if a planet has free oxygen in its atmosphere it can only be because some form of plant-like life is busily manufacturing it.

Whereas water is a requirement for life to survive; oxygen is a sign that life already exists on a planet.

The Earth
This view of the Blue Lagoon in Jamaica reveals the Earth to be a warm world with liquid water in abundance on which life flourishes.

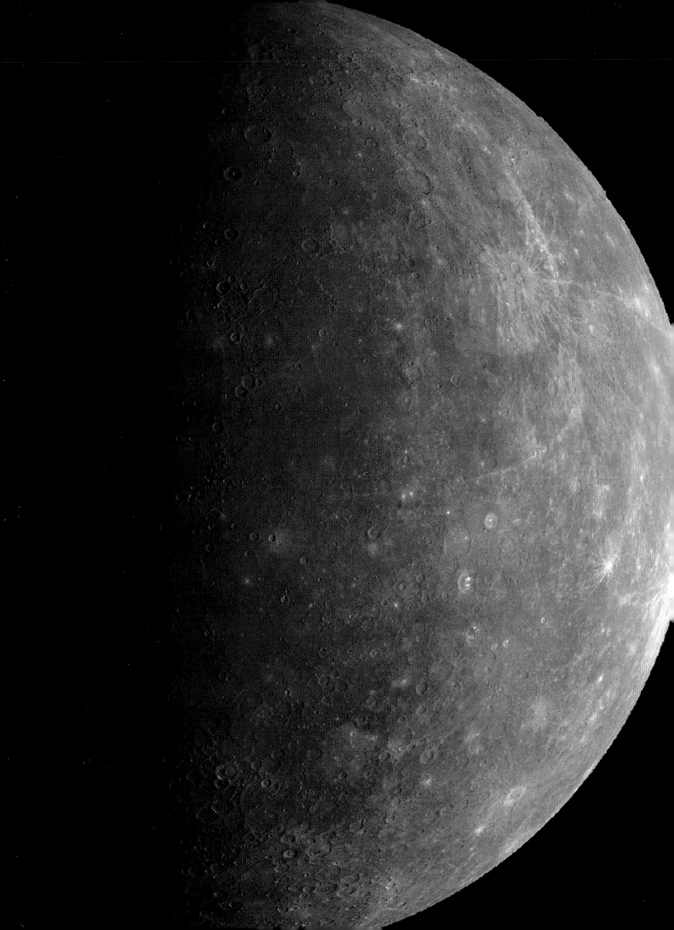

source of energy for life on other planets, but there is always the possibility of other energy sources to draw on.

Life forms that do make use of sunlight (or starlight) need to be on planets that are relatively close to their parent star. The Earth, as we shall see, occupies just such an orbit. To put the position of the Earth in the Solar System into perspective, it is convenient to start with the innermost planet, Mercury, and work our way out through the Solar System on a whistle-stop tour.

Distances across the Solar System are often measured in comparison with the average distance of the Earth from the Sun, which is defined as one astronomical unit (1 au). Mercury orbits the Sun at a distance of 0.39 au, taking just under eighty-eight days to complete one orbit. It is a small, rocky planet, with a diameter of 4,880 kilometres, making it intermediate in size between our Moon and the planet Mars. There are no traces of water on Mercury (indeed, it has no significant atmosphere), so we can definitely rule it out as a home for life as we know it.

Whereas Mercury has insufficient atmosphere to support life, the next planet out from the Sun, Venus, seems to have too much. This is particularly frustrating since, in terms of size and distance from the Sun, Venus is very nearly a twin of the Earth. It orbits the Sun at a distance of 0.72 au, taking 225 days to complete one orbit. It has a diameter of 12,104 kilometres (marginally less than the Earth) and a mass of 82 per cent that of the Earth. The atmosphere of Venus is an extremely dense layer of carbon dioxide, which produces a surface pressure ninety times the atmospheric pressure at sea level on Earth. The planet is swathed in clouds made of sulphuric acid droplets, from which acid rain falls but evaporates in the searing heat before it can reach the surface of the planet.

Partly because it is that bit closer to the Sun than the Earth, and partly because there is so much carbon dioxide in its atmosphere (which traps heat through the so-called greenhouse effect), the temperature at the surface of Venus soars above 450 degrees Celsius. This makes it impossible for liquid water to exist. Any water that ever did exist on Venus could only have been in the form of vapour, which would have risen high into the atmosphere and been broken down by ultraviolet radiation from the Sun into its principal components, hydrogen and oxygen. The hydrogen

Mercury
Mariner 10 visited the innermost planet in the Solar System in 1974, taking this picture of the battered planet as it approached. Airless and covered with craters, Mercury looks very much like the Moon.

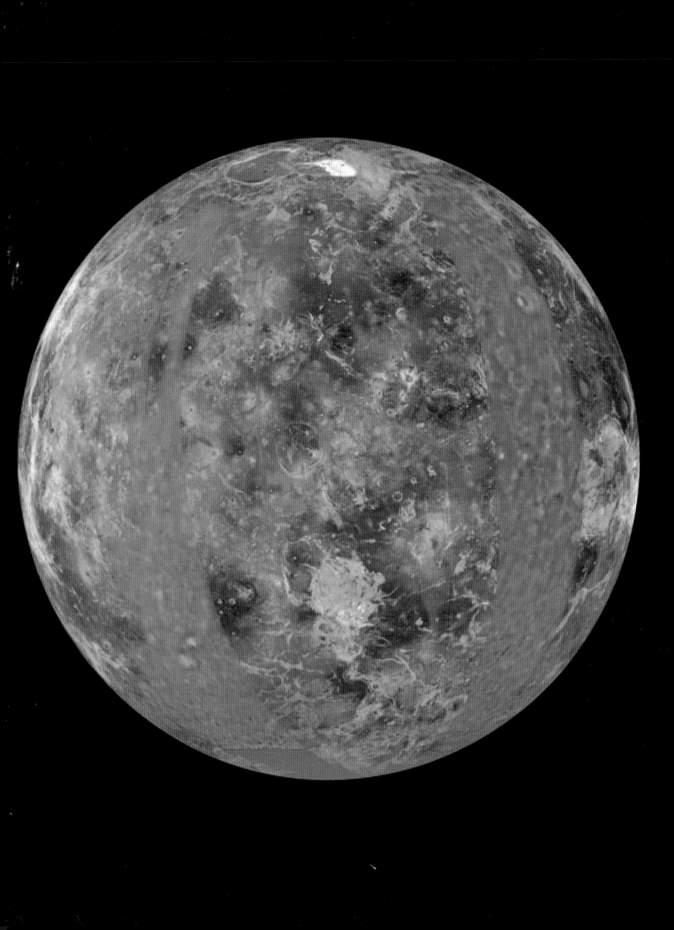

Venus
A false-colour image of
the surface of Venus
made from radar data
collected by the Magellan
Orbiter in the early
1990s. Since light cannot
penetrate the thick clouds
of Venus, the only way to
see the surface of this
planet is with radar.

(the lightest gas there is) would have escaped into space, while the oxygen would have become locked up in compounds such as carbon dioxide.

Earth became an ideal home for life (sometimes referred to as the Goldilocks planet, because, like baby bear's porridge, it is 'just right'). Being just a little bit further from the Sun (by definition, at a distance of 1 AU) than Venus, the corresponding reduction in solar radiation reaching the Earth allowed oceans of liquid water to form, which was good for life in itself. It also meant that carbon dioxide dissolved in these oceans and was laid down as carbonate rocks, thinning the atmosphere and weakening the greenhouse effect. Our home planet takes 365.26 days to orbit the Sun once. It has a diameter of 12,756 kilometres, and a mass of just under six thousand billion tonnes.

Although our Moon is both airless and lifeless (like Mercury) it has a diameter of 3,476 kilometres and a mass equivalent to 1.2 per cent the mass of the Earth.

The last of the small, rocky planets in our Solar System is Mars, which orbits the Sun at a distance varying between 1.38 au and 1.67 au

Man on the Moon
An astronaut drives a
Lunar Roving Vehicle
(Moon Buggy) over the
barren surface of our
closest companion,
the Moon, during the
Apollo 16 mission. So
far the Moon is the only
other body in the Solar
System that we have
visited in person.

Earth from space

A stunning view of the
Earth showing North and
South America alongside
the vast expanse of the
Pacific Ocean. Even from
space the Earth is
obviously a home for life.

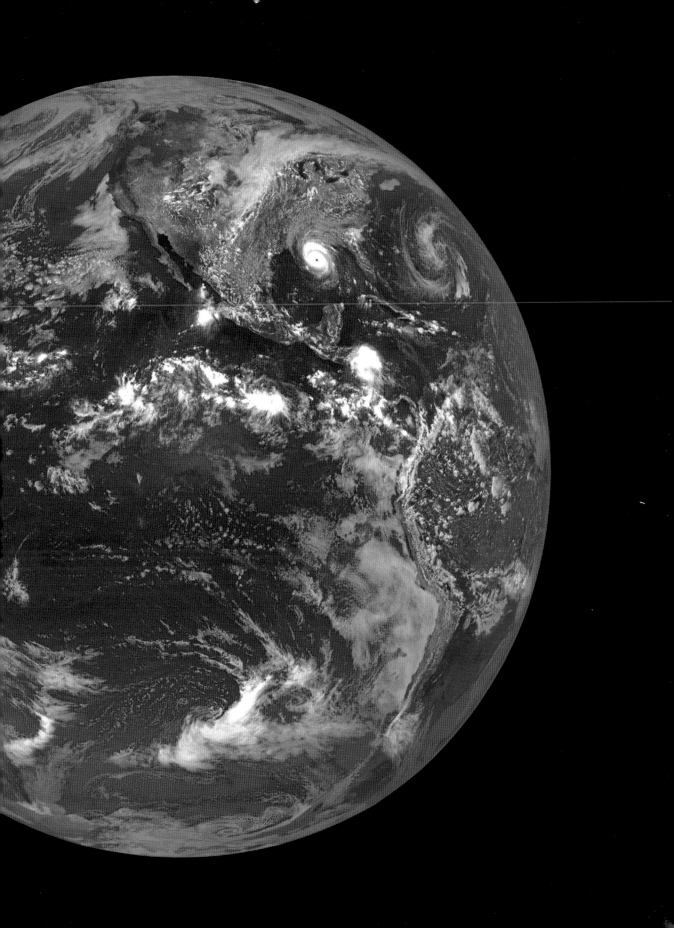

(because the orbit is elliptical), taking 686.98 days (nearly two Earth years) to orbit the Sun once. Although much smaller than the Earth, with a diameter of 6,795 kilometres and a mass of just over one-tenth the mass of the Earth (less than ten times the mass of the Moon), Mars does have a thin atmosphere, and there are clear signals on its surface that water once flowed there. The Martian atmosphere, such as it is, is mostly carbon dioxide, which is a very good greenhouse gas. However, it is so thin, and the planet is so far from the Sun, that even with the help of the greenhouse effect temperatures on Mars today dip below −100 degrees Celsius.

It seems that when Mars was young it must have had a much thicker atmosphere, made from gases spewed out by volcanoes, and this kept it warm enough for water to flow. But the gravitational pull of Mars is so weak (because it has so little mass) that most of this atmosphere has escaped into space, and at the same time the volcanoes have died out as the interior of the planet has cooled (smaller planets cool quicker than larger ones). The result is that the surface has gone into deep freeze,

Water tracts on Mars
The Mars Global Surveyor found many features on Mars that look like they were formed by liquid water flowing on the surface of this planet. Here a large valley system has been carved into the surface by the flow of some liquid which scientists think must have been water.

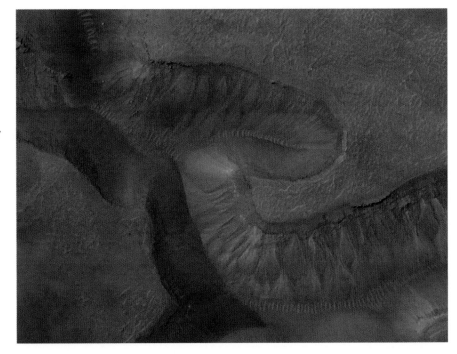

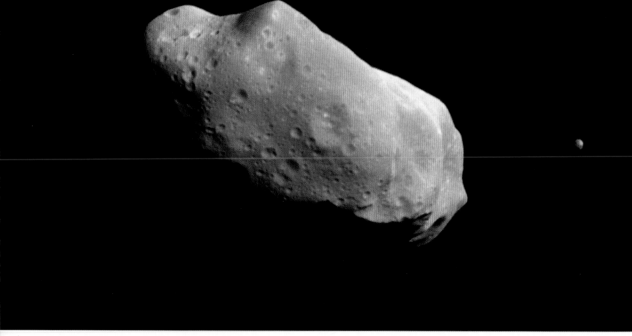

Ida and Dactyl
This close up of the asteroid Ida and its tiny companion Dactyl was taken by the Galileo probe en-route to Jupiter. Ida and Dactyl orbit just outside the main asteroid belt.

although there may still be plenty of frozen water just below the surface. There may once have been life on Mars, and just possibly cold-resistant life forms like bacteria may still survive today.

Beyond Mars, there is a region of space in which many lumps of cosmic rubble, known as asteroids (or, occasionally, as minor planets) orbit the Sun at distances ranging from 1.7 au to 4 au. There are at least one million objects, each a kilometre or more in diameter, in this asteroid belt, plus countless numbers of smaller pieces of rubble. But none of them have atmospheres, or water, and can be ruled out as homes for life.

Beyond the asteroid belt, there are four planets totally unlike any of the four inner, rocky planets. These are huge objects known as gas giants (see table on page 23). The gas giants themselves are not suitable homes for life as we know it, because (as their name suggests) they are largely balls of gas, with tiny solid cores buried deep beneath their clouds. But because they are so big, planets like these are relatively easy to detect and, as we shall see, the first direct evidence for planetary systems other than our own, around stars like the Sun, has come from detecting similar

planets to these. However, families of moons surround the gas giants in our Solar System, rather like a miniature Solar System. Some of these moons, rather than the gas giants themselves, could provide homes for life. One of the most exciting prospects is Europa (**see page 108**), a moon with a diameter of 3,138 kilometres which orbits around Jupiter, the largest of the gas giants in our Solar System. Europa has an atmosphere, and an ice-covered ocean of what seems to be liquid water. This is kept at temperatures above freezing, in spite of its great distance from the Sun, by the tidal forces exerted by Jupiter, which rhythmically squeeze and release the interior of the moon, thus generating heat.

Even the gas giants do not mark the boundary of the Solar System. Beyond Neptune, the most distant of the giant planets from our Sun, there is a belt of icy debris, reminiscent of the asteroid belt, which is called the Kuiper Belt. This is a ring of iceberg-like objects orbiting the Sun at distances beyond 50 au. Pluto, which for historical reasons is usually referred to as a planet, is actually just the biggest member of the Kuiper Belt, a lump of ice with about one-quarter as much the mass of

The Solar System, the distances, years, radii and masses of the nine major planets about the Sun.
The separation of planets into terrestrial planets (Mercury, Venus, Earth, Mars and Pluto) and gas giants, or Jovian planets (Jupiter, Saturn, Uranus and Neptune) is obvious in both size and mass.

Planet	Average distance from Sun (AU)	Orbital period (years)	Radius (km)	Mass (Earths)
Mercury	0.4	0.24	2440	0.06
Venus	0.7	0.62	6050	0.82
Earth	1.0	1.0	6380	1.00
Mars	1.5	1.9	3394	0.11
Jupiter	5.2	11.9	68700	318
Saturn	9.6	29.5	57550	95
Uranus	19.2	84.0	25050	14.5
Neptune	30.0	164.8	24700	17.2
Pluto	39.4	248.6	1700	0.002

our Moon. Further out still, literally halfway to the nearest star, there is a cloud of cometary material, known as the Oort Cloud, which surrounds the Solar System at distances between 50,000 and 100,000 au from the Sun. Beyond that lie the stars, and possibly other planetary systems that are homes for life.

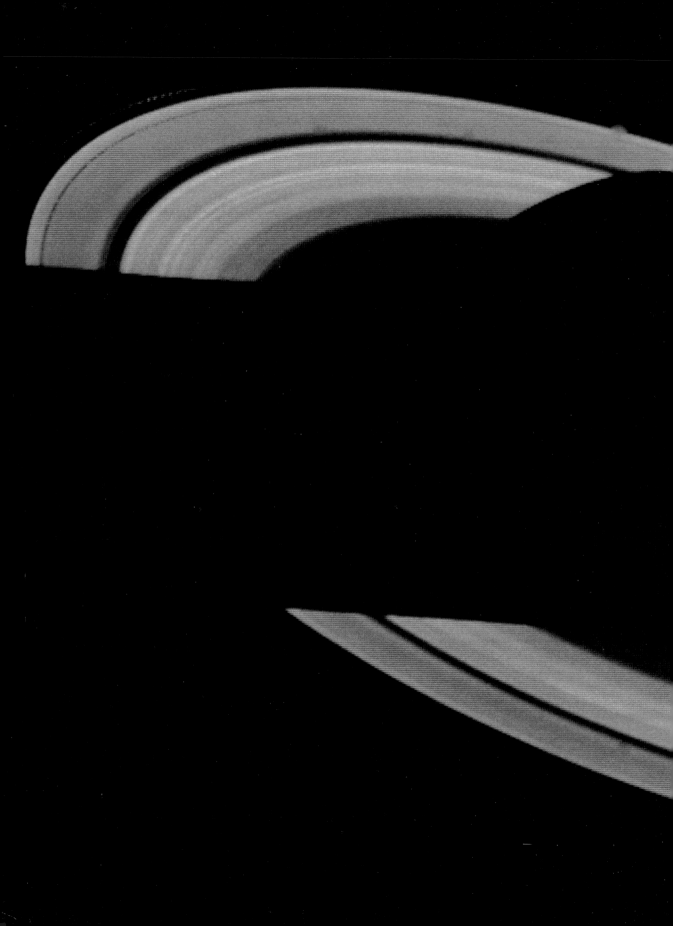

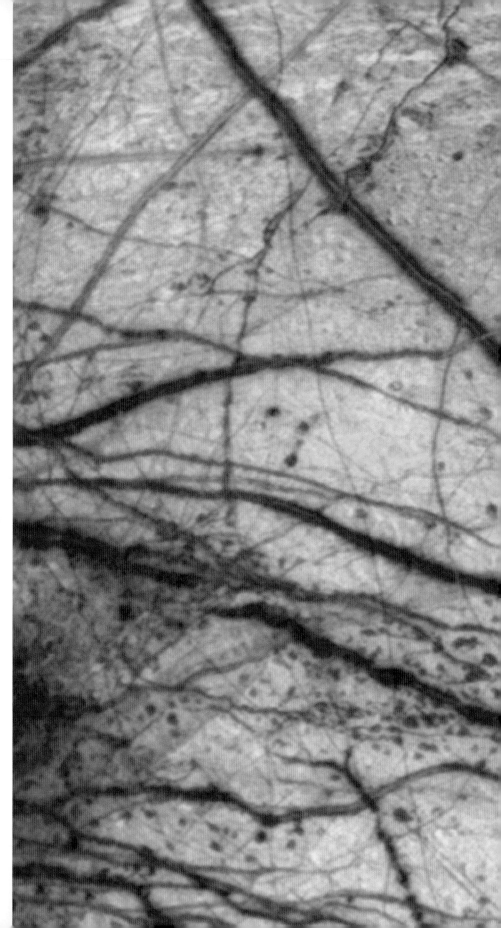

◄ **Saturn**
Voyager 1 captured this image of Saturn and its spectacular rings as it sped away from the planet in 1980. Sunlight striking the planet from the right is blocked by Saturn, which casts a shadow across the rings.

Ice on Europa ►
This close-up photograph from Galileo reveals the surface of Europa to be covered by a thick layer of frozen water. The lines that criss-cross the ice are cracks and fractures caused by impacts on Europa and the gravitational tug of Jupiter.

The five innermost planets (Mercury, Venus, Mars, Jupiter and Saturn) have been known since antiquity. They are all very bright and distinguish themselves by moving against the backdrop of the apparently fixed stars. Indeed, the word planet comes from the Greek for 'wanderer'.

It was thought that these were the only planets and nobody seriously considered the idea that the Sun may have more planets. That was until Sir William Herschel discovered Uranus on the night of Tuesday 13 March 1781, using a telescope he built himself, in the back garden of his house in Bath.

He observed the object over the course of several weeks and showed that the object moved compared to the background stars. He initially announced his discovery as a comet. Herschel was a music teacher and amateur astronomer at the time, although the word amateur hides the fact that he was, by far, the best telescope maker in the world.

When astronomers and mathematicians tried to fit the orbit of this new comet they found that it was almost circular, far beyond the orbit of Saturn. This meant that it was not a comet, but a new planet. Unfortunately, the technology to detect planets has moved on to the point that we cannot imagine anybody finding one from their back garden again (although you never know.).

As the years went by astronomers tried to fit the orbit of Uranus with models based on the gravitational pull of the Sun and small perturbations caused by the gravitational pull of both Jupiter and Saturn. They found that, no matter how hard they tried, Uranus did not follow the orbit they predicted. The best explanation was that an even more distant planet was perturbing Uranus.

In the 1840s John Couch Adams and Jean-Joseph Le Verrier worked independently to calculate the orbit of the unknown planet. Knowing its orbit would tell astronomers where and when to look for it, and in 1846 astronomers discovered what was to become known as Neptune.

One more planet (or pseudo planet) was still awaiting discovery. Again, the orbit of Neptune did not follow the predicted path. And again, another, more distant, new planet was postulated to account for this.

Clyde Tombaugh, an astronomer at the Lowell Observatory in the USA, discovered Pluto in 1929.

The Solar System was now complete.

Sir William Herschel (1738 – 1822)
The greatest astronomer of his day, Sir Willam Herschel discovered Uranus in 1781.

1.2 The Sun – our star

The Sun is an ordinary star. It looks so big and bright to us simply because it is so close – a mere 150 million kilometres away. Even the nearest stars (apart from the Sun) are several light years away; a light year is the distance that light can travel in one year, and for comparison the distance to the Sun is a little over eight light minutes. The diameter of the Sun is 1.4 million kilometres, a little more than one hundred times the diameter of the Earth. Because the volume of any spherical object is proportional to the cube of its diameter, this means that the Sun is about one million times bigger than the Earth. As the mass of the Sun is only equivalent to the total mass of about 300,000 Earths, its average density is just one-third that of the Earth, less than one-and-a-half times the density of water. That average density, though, covers a very wide range, from a superdense core 160 times the density of water (twelve times the density of solid lead) covering only 1.5 per cent of the volume of the Sun, to a very tenuous outer atmosphere. Sixty per cent of the mass of the Sun lies within less than 2 per cent of its volume.

Nuclear fusion. The heat and light upon which life on the surface of the Earth depends is all produced by nuclear fusion reactions going on in that superdense core. The Sun is made up of just over 70 per cent hydrogen by mass, slightly less than 28 per cent helium by mass, and approximately 2 per cent of other (heavier) elements. We know this from spectroscopy. Spectroscopy is the single most important tool of astronomy. It reveals the presence of different elements in the Sun and stars (and even in clouds of gas and dust in space) from their characteristic lines, as distinctive as a finger-print, which they produce in the rainbow spectrum of light (see box on page 37). Conditions in the core of the Sun are so extreme – in addition to the high densities, the temperature there reaches 15 million degrees Celsius – that hydrogen is converted into helium by nuclear fusion. This happens in several stages, starting with the collision of two protons (the nuclei of hydrogen atoms) and building up to alpha particles, the nuclei of helium atoms (see box on page 33). Each alpha particle has less mass than the sum of all of the protons that have gone into making it. The additional energy is

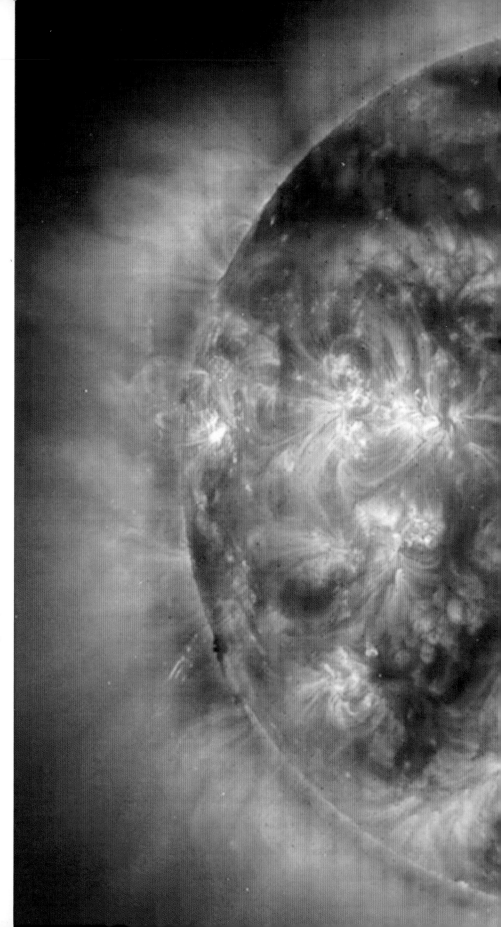

The Sun

This unusual image of
the Sun was taken by
the extreme ultraviolet
camera of the Solar
and Heliospheric
Observatory (SOHO).
It shows plumes of
hydrogen gas in the Sun's
atmosphere, which are
often associated with Sun
spots on the surface.
The quit surface of the
Sun we see in visible
light lies below this active
'atmosphere'

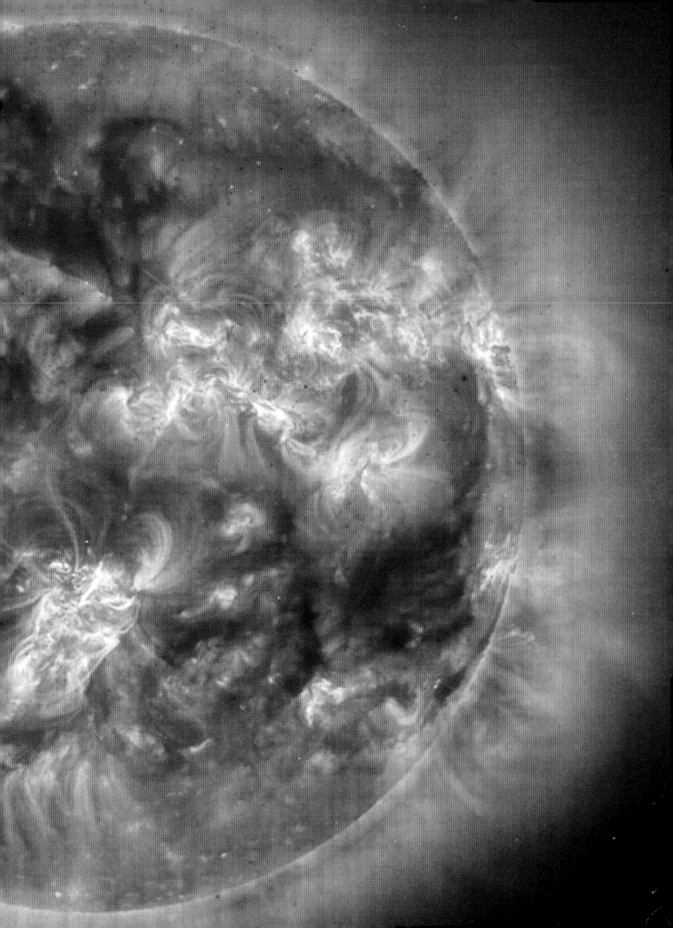

released in line with Einstein's famous equation $E = mc^2$ (to be precise, 0.7 per cent of the original mass is 'lost' when an alpha particle is made in this way). Nearly five million tonnes of mass is converted into pure energy in this way in the Sun's core every second. Even though the Sun has been 'burning' its hydrogen fuel for about 4.5 billion years, it started out with so much mass that it has only lost about 4 per cent of its original supply of fuel as yet. It is estimated that the Sun will keep shining in much the same way for several more billion years to come.

The energy released in this way is in the form of very highly energetic radiation, known as gamma rays. These gamma rays interact strongly with the charged particles that surround them, bouncing around as if they were in some kind of crazy pinball machine. Gradually, the radiation is degraded into X-rays, then visible light, and can be thought of as 'particles of light' or photons. If a photon could travel unimpeded (at the speed of light, of course) from the core of the Sun to its surface, it would take just 2.5 seconds to complete the journey; but because of this cosmic pinball effect, it actually takes about ten million years. When we look at the surface of the Sun today, what we see is a result of what was going on in its core ten million years ago. The fact that what is happening inside the Sun affects the surface so much later helps to average out any fluctuations, keeping a star like the Sun shining more or less steadily. This is good news for life forms on planets like the Earth which orbit stars like the Sun.

Star formation. A star forms from a fragment of an interstellar cloud, shrinking under its own weight. It is this shrinking that makes the interior of a star hot in the first place, because as the cloud gets smaller gravitational energy is converted into heat. If you drop a tennis ball from a tall building, it falls faster and faster until it hits the ground. Similarly, individual atoms in a collapsing cloud of gas, falling towards the centre of the cloud, move faster and faster until they hit other atoms. The temperature of the cloud (or anything else) depends on how fast the atoms and molecules it is made of are moving, and how hard they bounce off one another. Therefore the collapsing cloud gets hotter as it shrinks. Eventually, it is so hot that the pressure of all the atoms (strictly speaking, the nuclei) bouncing off one another stops it collapsing any more. For a

Nuclear fusion inside the Sun occurs in a three-step process, which converts four protons into one alpha particle. All atoms consist of a tiny core (the nucleus) made up of positively charged protons and neutral neutrons, surrounded by a cloud of negatively charged electrons (one electron for each proton in the nucleus). Hydrogen is the simplest atom, and consists of a single proton associated with a single electron. An alpha particle is equivalent to a nucleus of helium, and contains two protons and two neutrons. In a helium atom, such a nucleus is associated with two electrons.

Under the extreme conditions that exist in the core of the Sun (and other stars) electrons are stripped away from their nuclei, forming a mixture of positively charged nuclei and negatively charged electrons known as a plasma. Although the density in the core of the Sun is twelve times that of solid lead, the nuclei are so much smaller than atoms that even under these conditions they bounce around freely, colliding with one another.

This is identical to the way that atoms of a gas bounce around in the air that you breathe.

Energy generation inside the Sun begins when two protons get close enough to one another (in spite of their mutual positive charge, which tends to repel them from one another) to interact through quantum processes. This is an extremely rare event, which is why the Sun has burned so little of its nuclear fuel in the past 4.5 billion years. The first step is to form a nucleus known as deuterium, containing one proton and one neutron. This involves one of the original protons spitting out a positron (a positively charged counterpart to the electron) to get rid of its charge. Once this has happened a second proton can latch on to the deuterium nucleus to form a nucleus known as helium-3. Finally two helium-3 nuclei come together, ejecting two protons and leaving an alpha particle (also known as helium-4) behind. Each alpha particle contains two protons and two neutrons.

The p-p chain
Hydrogen is built up by actomic fusion into helium in the Sun, releasing the energy that powers the Sun and warms the Earth.

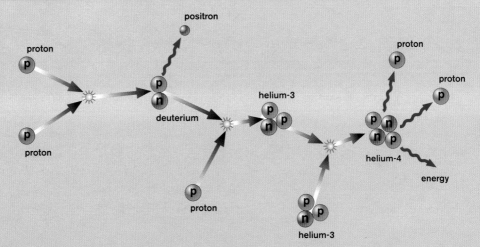

fragment of interstellar cloud as big as the one that formed our Sun, by the time this happens the temperature at the heart of the cloud has got so high that nuclear fusion reactions begin to take place. This releases energy and prevents the cloud from shrinking further providing there is a sufficient supply of nuclear fuel. The collapsed cloud has become a star. However, all of this only happens if something gives the original interstellar cloud a squeeze to start the ball rolling.

The Milky Way. Stars like the Sun form because processes occurring within our Milky Way galaxy squeeze clouds of gas and dust in space. All the stars we see in the sky with our unaided eyes are part of an island in space containing hundreds of billions of stars spread across a disc about 100,000 light years in diameter. This is known as the Milky Way, or the Galaxy (with a capital G); there are hundreds of billions of other galaxies in the Universe, but in this book we are only concerned with the Milky Way. Galaxies like our Milky Way are sometimes called spiral galaxies, because when seen face-on, they exhibit pronounced spiral patterns. The whole of this kind of Galaxy is rotating, but not as a solid object like a cd. Each star in the disc moves in its own orbit at its own speed, and a spiral pattern is produced where stars pile up temporarily in a kind of stellar traffic jam known as a shock wave. The shock wave moves around the Galaxy at a speed of about 30 kilometres per second, but the stars and clouds of material in the disc of the Galaxy move at speeds of up to 300 kilometres per second, overtaking the shock wave and passing through it. At the edge of the shock wave, clouds of gas and dust get squeezed and collapse. This collapse encourages the clouds to break-up and collapse further to make stars. The biggest fragments collapse to make the biggest stars. These stars burn their nuclear fuel very rapidly and end their lives in great explosions, known as supernovae, which send out blast waves that encourage more clouds to collapse and form more stars. Currently lying about two-thirds of the way out from the centre of the Galaxy at the edge of a spiral arm, our Sun and Solar System formed along with many other stars (and presumably many other planets) in just this way 4.5 billion years ago. However since then, we have travelled around the entire Galaxy about twenty times and lost all contact with our birthplace.

◄ **Spiral galaxy**
Very similar in structure to our own Milky Way, the spiral arms of this galaxy contain many young, bright stars. In the Milky Way the Sun lies about two-thirds of the way out from the centre.

A spectrum of light from the Sun

The dark lines that show up on this spectrum of sunlight occur where specific wavelengths of light have been blocked by atoms in the solar atmosphere. These lines allow astronomers to calculate the chemical composition of the Sun.

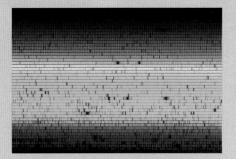

The cosmic bar code that is visible light can be passed through a triangular piece of glass, called a prism, and spread out to make all the colours of the rainbow. When physicists first examined in detail why sunlight spread out in this way in the nineteenth century, they noticed that the colours in the spectrum were crossed by many patterns of lines, which resemble the patterns of lines seen today in a bar code.

Studies in the laboratory then showed that each kind of atom produces its own pattern of spectral lines. Hence the patterns in sunlight are caused by the presence of different elements in the atmosphere of the Sun. All atoms of a particular element produce exactly the same pattern of lines, which can be bright lines if the atoms are hot and radiating energy, or dark if they are cool, and absorbing energy. For example, sodium atoms radiate bright orange lines if they are hot – which is why street lamps that contain a trace of sodium have their distinctive orange colour. The details of why each kind of atom

produces its own characteristic spectral bar code depend on quantum mechanics; but there is no need to understand quantum mechanics to read the bar code. Each time astronomers see a particular pattern of lines, they know a particular element is present – even if the light they are studying comes from a star dozens or hundreds of light years away.

The power of spectroscopy is highlighted by the way it was used to discover a 'new' element. There are hundreds of lines in the spectrum of light from the Sun, almost all of which were identified with known elements during the nineteenth century. But after eliminating all of these, astronomers were left with one set of unidentified lines. In 1868, the British astronomer Norman Lockyer said that these lines must correspond to a previously unknown element, and gave it the name helium from the Greek word for the Sun. Helium was only found on Earth in 1895.

It is not always simple to determine how much of each element is in a star from the lines you see in a spectrum. In the visible part of the Sun's spectrum there are many lines of iron. For a long time it was thought that the Sun consisted mostly of iron. However, a proper understanding of how gases at high temperature (plasmas) work shows that actually the Sun is virtually all hydrogen and helium with only 2 per cent attributable to other elements. It just so happens that the visible spectrum of the Sun has lots of iron lines and very few hydrogen and helium lines.

Making a spectrum

By passing white light through a prism, the different colours that make up this light are bent by different amounts, and spread to form a spectral rainbow.

sun

white light

prism

spectrum

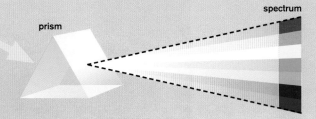

The Eagle Nebula

In this star forming
nebula, the bright light
from young stars reveals
pillars of dense gas
tipped with 'elephants'
trunks' where we believe
that new stars like the
Sun are forming. Each of
the tips of the elephants'
trunks in the top left of
the picture are actually
many times larger than
the Solar System.

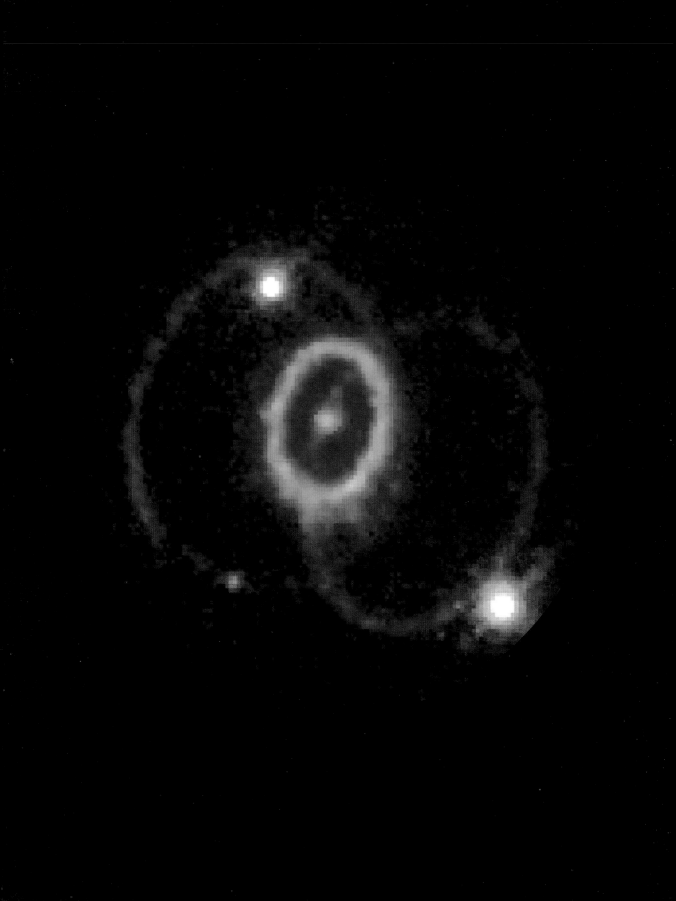

2

planets beyond

2.1 Stars, brown dwarfs and planets

Celestial objects can be roughly divided into three types: stars, brown dwarfs and planets (see page 44). Of these the only ones that are easy to see are stars that shine brightly like the Sun, producing prodigious amounts of light. Almost every object that is visible in the night sky is a star (with the obvious exception of some of the other planets in the Solar System).

The study of stars is the basis of most astronomy. The basic problem in astronomy is that the only way of finding anything out about things beyond our Solar System is by looking at the light that comes from them. The objects are too far away to actually go to, and so astronomers rely on a variety of sophisticated techniques to find out how big stars are and what they are made of. Considering that light is the only tool we have, it is amazing how much we know about the Universe and its constituents. Just as spectroscopy can tell us about the Sun, it can also give us a significant amount of information about other stars. But light is not just the visible light that our eyes detect.

The electromagnetic spectrum In 1800 the astronomer William Herschel discovered that there was more to light than the seven colours that we can see. He found that before red there is more light, invisible to our eyes, but detectable because it raised temperatures. Infrared radiation (literally 'before red') is light whose wavelength is longer than that of red light. Our eyes have not evolved to see lower than red or higher than violet, but the colours continue (many animals can see into the infrared and into the ultraviolet; plants that look dull to us may be highly coloured in the ultraviolet to attract insects).

We now know that the electromagnetic spectrum (of which visible light is only a tiny fraction) stretches from radio waves with wavelengths that can be measured in kilometres to gamma radiation with wavelengths smaller than an atom. The shorter the wavelength, the more energy it has simply because you are fitting more waves into the same distance. As a result, in one wavelength of a radio wave you can have billions of wavelengths of visible light. This is very useful as it means the hotter (more energetic) something is, the shorter the wavelength of light it emits.

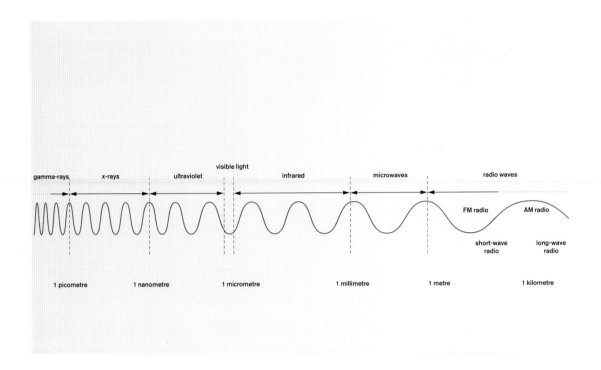

gamma-rays   x-rays   ultraviolet   visible light   infrared   microwaves   radio waves

                                                                                                            FM radio   AM radio

                                                                                                        short-wave   long-wave
                                                                                                          radio   radio

1 picometre   1 nanometre   1 micrometre   1 millimetre   1 metre   1 kilometre

The EM spectrum
The light we see is only a tiny fraction of the a huge range of electromagnetic (EM) radiation. The longest radio waves are over 100 kilometres long, while the shortest gamma rays are less than one billionth of a millimetre in wavelength.

We all know this from everyday life. A lump of metal when heated will feel hot before it begins to glow (your skin sensing the infrared radiation). Then, as it is heated more it will start to glow red, then yellow then blue – its colour changing as the energy of the light it emits increases (and so the wavelength gets shorter). The same is true for stars. This tells us how hot a star is – red stars are cool, yellow stars hotter and blue stars hotter still.

Astronomers had the ability to determine the temperature of a star as well as its composition. In the 1930s and 1940s, an understanding of how stars work was achieved – all this from a little bit of light collected in telescopes!

Stars balance the inward pull of gravity with the outward push of pressure caused by nuclear fusion. A very large star has a greater gravitational force than a smaller star, so the pressure pushing outwards must be correspondingly larger. For higher pressures read higher temperatures, and so the mass of a star can be deduced from its temperature.

Astronomers classify stars by their temperature (which is equivalent to their mass) in the Harvard sequence: O B A F G K M (easy to remember using the rather old-fashioned mnemonic Oh Be A Fine Girl, Kiss Me). At the high temperature end of this sequence are the massive O and B stars larger than about eight times the mass of the Sun, and at the low temperature end are K and M red dwarfs, far smaller than the Sun. The Sun itself sits near the middle – it is a G-type star.

Stars have a range of masses from about one-tenth to about 100 times the mass of the Sun. The larger a star is the hotter and denser is its core and the faster it burns its nuclear fuel from hydrogen to helium. What this means is that (rather counter-intuitively) the larger a star is, the shorter its life. A star one hundred times larger than the Sun will burn all of its hydrogen fuel in only a few million years (after burning hydrogen a rather complex, but short lived series of events occur which end in the star blowing itself up – but more of this later). A star the size of the Sun will live for about ten billion years before exhausting its hydrogen and will die more quietly, and a star about one-tenth of the Sun's mass might live for many trillions of years. The rate at which stars use up their fuel also determines how bright they are. Massive stars using their hydrogen at a

Relative masses
Planets and stars cover a huge range of masses. Planets are generally less than 0.001 the mass of the Sun. A celestial object that is larger than about one-tenth the mass of the Sun is termed a star. Anything in between is called a brown dwarf.

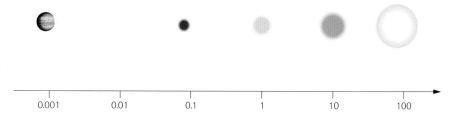

| 0.001 | 0.01 | 0.1 | 1 | 10 | 100 |

prodigious rate will shine very brightly through this energy release making them very easy to see from vast distances. The smaller a star, the fainter and more difficult it is to see.

The Eddington Limit, derived by the great British astronomer Sir Arthur Eddington sets the upper limit of one hundred times the mass of the Sun (or 100 solar masses). Stars heavier than this limit are blown apart by their pressure; their cores become so hot that nuclear fusion produces too much heat and light for the star to contain and it will be destroyed.

Most stars are rather dull red dwarfs, much smaller than the Sun. The number of stars with a given mass falls as the mass increases. This fall is

Betelgeuse
On the left, forming Orion's left shoulder in the night sky, Betelgeuse is a massive star 800 times the diameter of the Sun.

In the right-hand image a small blob can be seen slightly to the right of the star Gilese 623. This blob is Gilese 623b, a star so small that it is only just large enough to burn hydrogen.

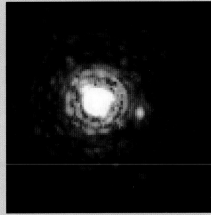

Astronomical objects are generally classified into three categories depending on their mass and properties: stars, brown dwarfs and planets.

Stars: stars are large balls of gas (virtually all hydrogen and helium) that are massive enough so that the temperatures and pressures in their centres are high enough to maintain nuclear fusion as happens in the Sun (see Section 1.2). It turns out that anything larger than about one-tenth the mass of the Sun is able to keep nuclear reactions going. The fusion of hydrogen atoms into helium atoms creates energy and enables stars to shine by their own light.

Brown dwarfs: brown dwarfs are an intermediate stage between planets and stars. If a collapsing gas cloud creates an object not quite big enough to start nuclear fusion, but too big to be a planet then it is known as a brown dwarf. Brown dwarfs do not fuse hydrogen into helium and so do not shine by generated energy, but they do trap a lot of heat during formation and can glow faintly as they slowly leak this heat out into space.

The largest brown dwarfs are just too small to be stars and so have a mass a little less than one-tenth the mass of the Sun. The dividing line between planets and brown dwarfs is far less distinct. We like the definition that a brown dwarf is too small to burn hydrogen but is capable of a short period of deuterium burning.

Deuterium is a heavy form of hydrogen; the vast majority of hydrogen in the Universe has only one proton as its nucleus but deuterium has one proton and one neutron making it twice as heavy as normal hydrogen. Deuterium replaces hydrogen in 'heavy water' making it weigh more than normal water. Deuterium is much easier to fuse than normal hydrogen but since it is so rare (only one deuterium atom in one million hydrogen atoms) the fuel is exhausted quickly and very little energy is generated. By this criterion, the smallest brown dwarfs are about ten times larger than Jupiter (or one-hundredth the mass of the Sun); anything smaller cannot produce its own energy at any time.

Planets: planets, as you might expect, are smaller than a brown dwarf – anything smaller than about ten times the mass of Jupiter. We like the definition of brown dwarfs stated above because Jupiter is classed as a planet (as most of us think of it, although some astronomers do like to class it as a brown dwarf). Planets then range from the large gas giants to the small rocky worlds like Earth.

There is another class of objects – minor planets such as asteroids and comets. These objects are so small (generally the size of large mountains) that they are very difficult to find, even when relatively close to the Earth.

roughly proportional to the square of the mass, so there are about five times fewer stars of two solar masses than of the mass of the Sun. This means that massive O- and B-type stars are very rare indeed, while the Galaxy consists almost entirely of M- and K-type stars.

Where should we look for planets? The only example we originally had of a star with planets around it was our own Sun, so the obvious place to look for planets is around stars similar to the Sun. This is not just a case of solar-centricism. We mentioned that the lifetime of a star depended on its mass. Massive stars just do not live long enough to form planets. Any star larger than about four solar masses lives for less than half a billion years – this is the length of time we think it took to form the planets in the Solar System. This means the largest stars we should search for planets are F-type stars; anything smaller may well have planets.

Most stars are part of a binary system – that is, they form in pairs and remain together. About two-thirds of all individual stars are part of a binary system. The problem that this causes in the search for planets is that astronomers believe that stars forming in a binary are unlikely to form planets, as the companion star disrupts the planet-forming disc. Even if planets do form the strong gravitational interactions of the two stars will throw any planets out of the system.

Our starting restriction for planet searching then is individual stars of type F or smaller – this leaves about 30 per cent of all stars as good places to start looking for planets. Given that the Milky Way contains about 300 billion stars this gives us a respectable 100 billion reasonable possibilities.

Most of these stars are too far away to be of any use, so we must look at stars that are relatively nearby. As most people are aware astronomical distances are generally measured in light years (actually, astronomers prefer to use parsecs where one parsec is about 3.2 light years but since parsecs are not familiar to most people we will stick with light years). A light year is the distance travelled by light in one year and considering that light travels at 300 million metres per second and circles the Earth seven-and-a-half times in that second, it's a very long way. A light year totals about 9.6 trillion kilometres! Numbers like these mean nothing to astronomers either which is why new distance measurements like the light year were invented. In order to be comprehensible we want the unit of measurement to normally be about one. A star being ten light years away and weighing three solar masses is far more comprehensible than it being ninety six trillion kilometres away and weighing six thousand trillion trillion tonnes.

There is roughly one star in every thirty cubic light years of space (meaning that, on average, stars are about three light years away from their nearest neighbour). Out to a distance of about 100 light years there are about 100,000 stars, and this is the sort of sample we have to examine in detail to search for planets. Only 30 per cent of these stars are good places to look, but it leaves a set of about 30,000 stars to search.

2.2 How to find planets

Planets are very difficult to find. The problem is that planets do not provide any light of their own, they just reflect light from their star. As planets are small compared to a star, the amount of light they reflect is

The Mira binary system
Located 250 light years away, the two stars in the Mira binary system are located only 40 au apart. The Hubble Space Telescope was able to seperate the two stars, and close-ups of Mira A (bottom left) and Mira B (bottom right) show how the gravitational attraction between the two stars is dragging material from both of them.

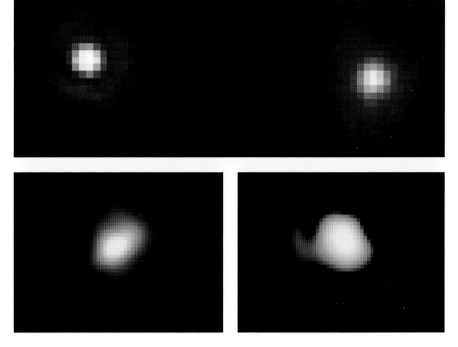

generally less than about one-billionth of the light coming from the star. As we will mention later, it is just possible to find planets by looking for them directly, but this is very difficult.

Since looking for the light reflected from planets is so difficult astronomers started looking for more indirect ways of detecting extrasolar planets. We think of the Earth orbiting the Sun, moving around while the Sun stays still. This is not exactly right. The Earth and Sun both move around their 'centre of mass' (see box page 49).

Thinking only of the Sun and Earth for the moment, the Sun is so much bigger than the Earth that the centre of mass is near the centre of the Sun. As far as we are normally concerned, the centre of mass 'is' the centre of the Sun, and the Sun does not move. But this is not exactly true – the centre of mass of the Earth-Sun system is not quite at the centre of the Sun, which means that the Sun moves a very small amount. The amount of time taken for the Sun to travel around the centre of mass is the same amount of time taken for the Earth to travel around the centre of mass – one year.

In fact, stars always take the same amount of time to travel around the centre of mass as the planet does. Jupiter lies five times further away from the Sun than the Earth and takes about twelve years to complete an orbit. Jupiter is by far the largest planet in the Solar System with a mass about one-thousandth of the Sun. This means that the centre of mass of the Sun-Jupiter system is about 800,000 kilometres from the centre of the Sun (this is slightly more than half the diameter of the Sun). The Sun orbits around this point every twelve years at an average speed of about 12.5 metres per second. In reality, the Sun follows a complicated dance around the centre of mass of the entire Solar System, affected by the gravitational force of all the planets. But Jupiter, as the largest planet, is by far the greatest influence.

The important thing about these numbers is that, on an astronomical scale, they are very small. In astronomy we talk of distances in, at the very least hundreds of millions of kilometres and speeds of kilometres per second. The distance the Sun is shifted is only about twice the distance to the Moon – a distance people have travelled. The speed at which the Sun moves due to Jupiter's gravitational force is only slightly faster than a world-class 100-metre runner. As you can probably guess, in a subject

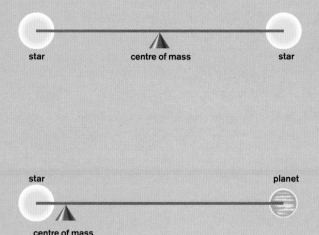

star centre of mass star

The centre of mass is the point along the line joining two objects where the ratio of the masses and distances are the same. Imagine a stick with weights at either end. Where you can put your finger to balance the weights is the centre of mass. If the two weights are equal the centre of mass will be in the middle of the stick (as in the upper diagram). The larger one of the weights is relative to the other, the closer the centre of mass is to the bigger weight (as in the lower diagram). Generall stars are more than 1,000 times heavier than planets so the centre of mass is very close to the centre of the star – but crucially not at the centre.

star planet

centre of mass

used to looking for big things, such small perturbations are very difficult to see. There are two simple things we could do to increase the wobble of the Sun caused by Jupiter which would make it easier to see. The most obvious is to make Jupiter bigger – the more massive the planet, the further from the centre of the star the centre of mass is. The second possibility would be to move Jupiter closer to the Sun, this would make Jupiter's orbital period (year) shorter and so increase the speed at which the Sun would orbit the centre of mass. By looking at the spectrum of a star it is possible to work out its type (O B A F G K or M) and from that its mass. Knowing the mass of the star the mass of the planet causing any wobble is easy to find.

Astrometry To find an extrasolar planet from the wobble it induces in the parent star we need to consider the angle at which we are looking at the planetary system. If we were to sit outside our own Solar System then the effect we would see would depend on the inclination of the orbits of the planets to our line of sight.

The two extremes are looking at a planetary system face-on or side-on. Face-on is like looking at the Solar System from above one of the Sun's poles; from here we would see the Sun move in a little circle around the sky as time passes due to the pulling of Jupiter. Side-on we would see the Sun simply move from one side to another as Jupiter travels around it.

The simplest thing to do would seem to be to look for a wobble in a star's motion as it travels through space. This may sound like a very easy job, but it is fraught with difficulty due to two problems. The first problem is the atmosphere of the Earth. One of the most beautiful features of stars is the way they twinkle in the sky, but this twinkle is the bane of an astronomer's life. Stars twinkle because the path of light through the atmosphere is bent by the air. The atmosphere is in constant motion, due to turbulence caused by convection and winds blowing at different altitudes moving air and slightly changing the amount of air between you and a star all the time. This constantly changes the exact position of the image of a star in the sky, which we see as twinkling. If you want to know 'exactly' where a star is, then its apparent position in the sky changing all the time is not

Planetary inclination
The left hand image shows a planetary system viewed side-on. The image to the right shows an identical planetary system viewed face-on. Such differences in inclination change the way we perceive the gravitational pull of a planet on a star and alters the technique needed to discover a planet in that system.

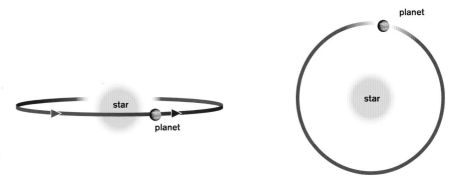

helpful. To help avoid this problem, astronomers put telescopes on the tops of high mountains, above as much of the atmosphere as possible. Even better is putting telescopes into space, like the Hubble Space Telescope, where there is no atmosphere to mess up your observations. But this is not the end of the problems. Telescopes are not perfect and there is a minimum resolution that they can achieve. The optics in telescopes smear light into a small circle and cannot focus it exactly.

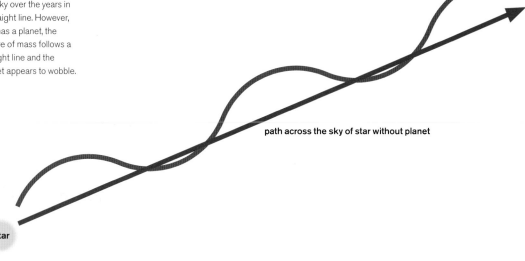

Astrometry
When a star has no planets, it moves across the sky over the years in a straight line. However, if it has a planet, the centre of mass follows a straight line and the planet appears to wobble.

path across the sky of star with planet

path across the sky of star without planet

star

Double vision
The problem caused to astronomers by Earth's atmosphere can be seen by comparing these two images of the same part of the sky. The blurred image on the left is a view from Earth, while the image to the right is a picture taken above the atmosphere by the Hubble Space Telescope.

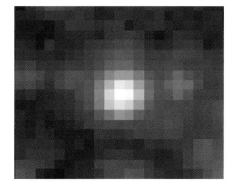
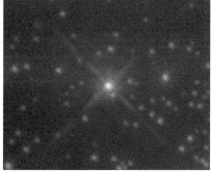

These 'seeing' limits (how small an object it is possible to see) affect all observational astronomy. For planet-hunters, though, they are everything. Planet-hunters are looking for very small changes in a star's position due to the influence of planets. Detecting these changes are at the very limit of astronomical technology.

As we have seen, the change in the position of a star due to a planet is very small. Astronomers normally measure the size of objects in the sky in terms of angles. This is because the distances to many astronomical objects are not known absolutely. When the Sun moves due to the influence of Jupiter, the size of that movement depends on how far away the observer is when the shift is observed.

Very important to astronomy is the idea of angular size. This is the size that something appears to us as we look at it from the Earth. The basic astronomical angular size measurement is the arcsecond. If you take a penny (about one centimetre across) and hold it 60 centimetres away its angular size will be one degree. At sixty times further away (34 meters) it will be an arcminute across and sixty times further still at two kilometres down the road it will be an arcsecond in size. (This rather odd division of the circle comes to us from the Babylonians.) The size of wobbles that we are looking for from a Jupiter-mass planet around a nearby star are measured in milliarcseconds, the distance across a penny at 2,000 kilometres!

Despite these problems astrometry has a long and distinguished history. In 1844 Freidrich Wilhelm Bessel wrote to William Herschel (about

Sirius A and B
Until recently, Sirius B (seen as a small blob of light to the top right of Sirius A) was invisible. It was first discovered by astronomy by the wobble it caused in the motion of Sirius A.

whom we have already heard) to tell him that the positions of the stars Sirius (the brightest star in the sky) and Procyon changed their positions in such a way that it implied they had invisible companions. In 1862 and 1896 respectively observations of the stars showed that both stars had neighbours. These neighbours were far too massive to be planets (or even brown dwarfs), but it was a very early confirmation that astrometry could detect hidden companions.

The history of planet detection by astrometry is less impressive. In the 1960s it was announced that Barnard's Star, the second closest star to the Sun (at only six light years) had a planetary companion. Observations of the wobble of Barnard's Star in the 1960s apparently showed that the star wobbled by a massive 24.5 milliarcseconds due to the influence of a planet about the same mass as Jupiter. Unfortunately further observations showed that it was a flaw in the telescope that was being used that caused the apparent motion, not a planet after all.

Despite several claims, astrometry failed to find any extrasolar planets. In the late 1980s, however, another technology had improved sufficiently to join the hunt for planets.

Doppler spectroscopy If a star is moving the positions of the lines in its spectrum are moved due to the Doppler shift (see box page 55). The amount that the lines have shifted tells us how fast and in which direction (towards or away from us) the star is travelling. Doppler methods have been used for most of the past one hundred years to discover the structure of our Galaxy and even that the Universe is expanding.

In the late 1980s it became possible to look for the shifts in the spectra of stars caused by their movement due to a massive planetary companion. In the case of the Sun looked at side-on it would move towards you for six years with a maximum speed of 12.5 metres per second and then move away from you for six years, reaching the same speed as it moved in the opposite direction.

Stars and galaxies move with speeds measured in kilometres per second (often hundreds of kilometres per second). These high speeds cause a large Doppler shift of the spectrum and are fairly easy to measure. The reason that Doppler spectrometry could not be used to find planets was that the uncertainty in Doppler shifts was far larger than the few tens of metres per second wobble that planets cause. The change in the position of a spectral line in the Sun due to Jupiter is a mere 1/25 millionth of the wavelength of the line, a wavelength already measured as ten-millionths of a metre!

The standard method of determining Doppler shifts is to observe a star for a while, passing the light from the star through a prism to create a spectrum. To work out what the rest positions of lines should be, every so often the telescope should be used to look at an electric lamp in the observatory (whose speed relative to us is zero and whose lines are known). The two spectra can be checked against lines whose positions we know and the shift in the lines of the star can be calibrated.

The main source of error in the Doppler shift is caused by the swapping of the target to and from the star and lamp. This can be avoided by looking at the star through a gas in the observatory. The gas then produces calibration lines in the same spectrograph. This means that the spectrum is calibrated automatically but has the disadvantage that the spectrum of the star is full of lines that do not belong to it. These additional lines means that the spectograph can cause more problems than it solves.

One of the earliest gases that planet-hunters used was hydrogen fluoride – a very poisonous gas. So poisonous in fact that many other astronomers refused to stay in the observatory while the spectograph was being used! In the late 1980s, however, two main groups had started observing stars with spectrometers capable of detecting wobbles as small as a few metres per second – easily good enough to detect the influence of Jupiter around the Sun. Even good enough to detect the Doppler shifts of runners in the Olympic 100 metre finals! Confidence was high that if planets were out there, they would soon be found. And so they were, but in the most unusual place.

2.3 Pulsar planets

The first planets to be discovered outside our Solar System were found as long ago as 1991. But these three planets are completely unsuitable as homes for life as we know it, and the star they orbit is very different from

The Doppler shift is an effect everybody knows. When a fire engine is travelling towards you, the sound of its siren is higher pitched than usual, and as it passes the pitch drops. This is because the sound waves coming from the siren are 'bunched-up' as the fire engine catches-up with the sound; as it travels away the sound waves are lengthened. The same effect occurs with light waves. If something is moving away from us the light is stretched out and becomes redder (the famous 'red shift'), and if something is moving towards us the light waves become shorter and the light becomes bluer. This happens all the time, but we never see it in everyday life because an object needs to be travelling a reasonable fraction of the speed of light for the effect to be noticeable to the human eye.

Doppler shifts
The top star is stationary and its spectrum shows a line at a place we see in a laboratory on Earth. If the star is moving the shape of the line alters, and becomes redder if the star is moving away from us and bluer if it is moving towards us.

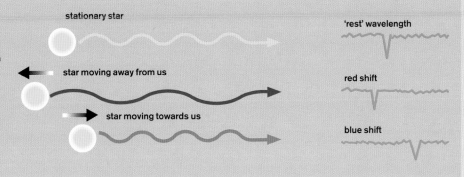

stationary star

star moving away from us

star moving towards us

'rest' wavelength

red shift

blue shift

our Sun. Even so, the nature of these planets and their orbits provide some intriguing insights into how planets form, and the chances of finding other planets similar to Earth.

The most important point that arises from this discovery is simply the fact that these planets exist. This proves that planets can form around other stars, and that the Solar System is not unique – something nobody could be absolutely sure of before 1991. However the strangest thing about these planets is that they orbit a stellar remnant called a neutron star (**see box page 61**), and astronomers are sure that the processes which make neutron stars would release any planets that already orbited the star. When a neutron star forms, it loses so much mass that its gravitational grip on its family of planets is weakened, enabling them to escape from their orbits.

Neutron stars are very compact objects, with about the same mass as our Sun packed into a sphere only a few kilometres across. They do not generate heat in their interiors, and do not shine, so they cannot be found using traditional optical telescopes. Fortunately many neutron stars generate intense bursts of radio noise, and can be detected using radio

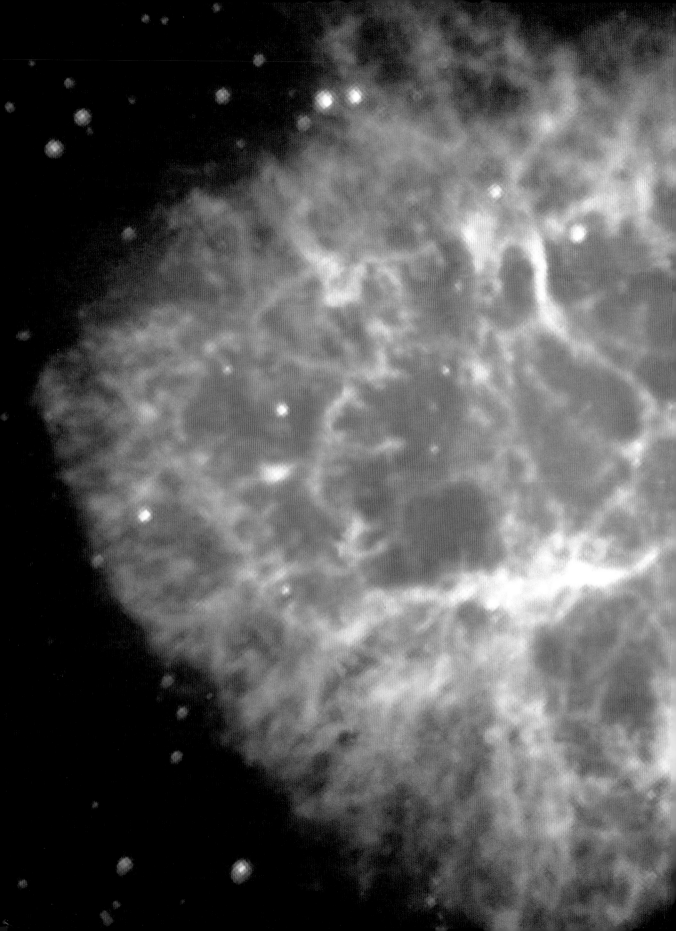

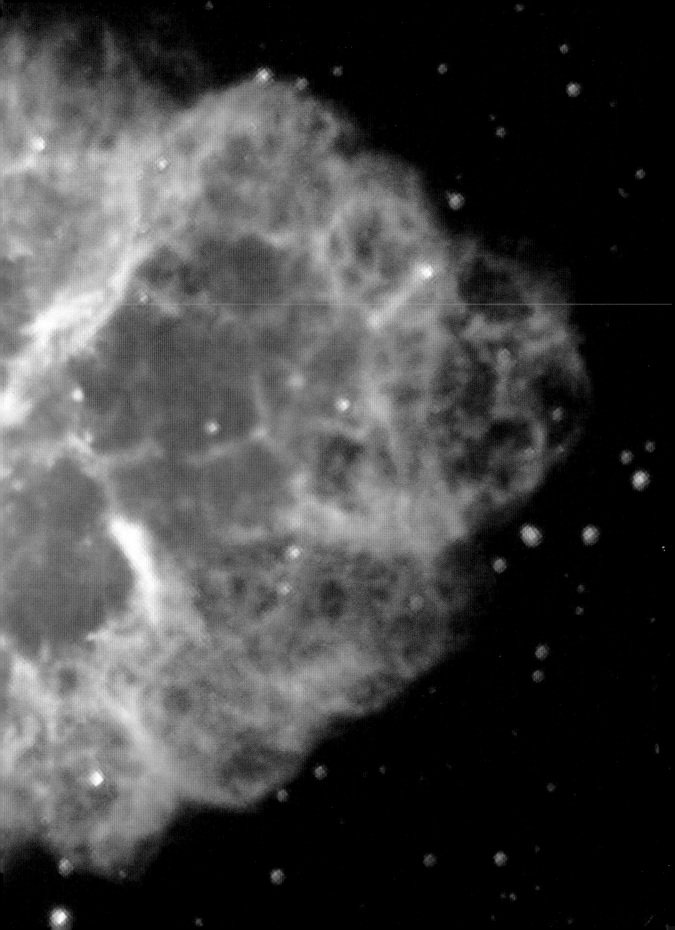

telescopes. The radio noise is beamed out from the spinning neutron star rather like the way light is beamed out from a lighthouse, and as the neutron star spins the radio beams flick past our radio telescopes, producing regular pulses of noise which give these objects their name – pulsars.

All pulsars spin at a very regular rate, so the pulses arrive with regularity as reliable as the finest clocks on Earth (some are even more reliable than our clocks). But they all spin at different rates, some as slowly as once every second and some as fast as once every few milliseconds. It was one of these millisecond pulsars, an object prosaically known as PSR B1257+12, which was found to have three planets in orbit around it.

The 1991 discovery of extrasolar planets was made using very subtle measurements of apparent changes in the speed with which the pulsar 'clock' is running. These rhythmic changes can be explained in terms of a regularly varying Doppler effect (**see page 55**), caused by the combined influence of the three planets tugging the neutron star to and fro. There is enough information available from these studies to calculate both the masses of the planets and their distances from the neutron star, and these properties are intriguingly similar to the properties of the three innermost planets in our Solar System. The three pulsar planets have masses equal to approximately 2.8 times, 3.4 times, and 1.5 per cent the mass of the Earth. The distance from the central star of their orbits, relative to the Earth-Sun distance are 0.47 au, 0.36 au and 0.19 au respectively. This means that the time it takes for them to orbit the pulsar once (their 'year') is 98 days (for the outermost), 67 days (for the middle planet) and 25 days (for the innermost), respectively.

The pulsar planets are a little closer to their parent star than the inner planets of our Solar System are to our Sun. Despite this, the overall pattern immediately reminds us of the inner Solar System with two relatively large planets (equivalent to Earth and Venus) orbiting beyond a much smaller planet (equivalent to Mercury). But the comparison is even more striking when you look at the numbers. The mass relative to the Earth (taking the mass of the Earth as 100 per cent) of Venus is 82 per cent, and the mass of Mercury is 5.5 per cent. The actual spacing of the orbits in our Solar System is in the ratio1:0.72:0.39, while the spacing of the three pulsar planets is in the ratio 1:0.77:0.4.

◀ **The Crab Nebula**
The Crab Nebula is the result of a supernova explosion that was observed 900 years ago. In the middle of the picture is the Crab Pulsar, whose structure is similar to the pulsars around which planets have been found.

Titius-Bode Law This tantalizing similarity in the spacing of the orbits has revived interest in an old idea, known as the Titius-Bode law, derived from comparison of the orbits of the planets in our Solar System. This 'law' was first noticed in the eighteenth century, initially by the German Johann Titius but popularised by his compatriot Johann Bode. At that time, only six planets were known – Mercury, Venus, Earth, Mars, Jupiter and Saturn. Titius and Bode pointed out that there is regularity in the spacing of their orbits, which can be summarized like this.

Take the numbers 0, 3, 6, 12, 24 and so on, with each number after the first two double the preceding number. Now add 4 to each number to give the sequence 4, 7, 10, 16, 28 and so on. If you set the distance from the Earth to the Sun as 10 units, this numerical sequence gives the correct orbital distances for Mercury (4 units), Venus (7 units) and Mars (16 units). Then there is a gap in the sequence, with Jupiter at 52 units and Saturn at 100 units, exactly where the law 'predicts'. When Uranus was discovered in 1781, orbiting the Sun at a distance of 196 units, not far from the orbit predicted by the law, it seemed that Titius and Bode had discovered a fundamental truth. Even the gap between Mars and Jupiter was filled in during the nineteenth century, when the asteroid belt was discovered. Interest in the law waned, however, when the orbits of Neptune and Pluto turned out not to match its predictions.

However the orbits of the pulsar planets do closely match the Titius-Bode law – just multiply the numbers 1, 0.77 and 0.4 by 10 and you get the ratios 10:8:4. Perhaps, after all, the law is telling us something about how planets form near to a star with about the same mass as our Sun. This is because the only explanation for the existence of these planets in the orbits they are found in is that they did indeed form there after the pulsar had formed.

This is not just a guess. Millisecond pulsars can only spin so fast if they have been accreting material. Material falling onto a spinning neutron star will make it spin faster, just as the Sun was made to spin faster, like a skater pulling in their arms, when the Solar System formed from a shrinking cloud of material. That material may have been debris from a companion star disrupted in the supernova explosion that gave birth to the pulsar, or from a cloud of material in space – it does not matter. The

The Hourglass Nebula
The final fate of most stars (including the Sun) is to become a planetary nebula like this one. The pinching together of the nebula in the middle may be due to a disc or planetary system around the star.

When a star like the Sun ages and uses up the nuclear fuel in its core (which will not happen to the Sun itself for about five billion years), it swells up and puffs away material into space, forming a cloud of gas called a planetary nebula. Then, the inner core of the star settles down into a single lump of matter, a stellar cinder, containing a little less mass than the star started out with, in a sphere about as big as the Earth. This stellar cinder is called a white dwarf. One cubic centimetre of white dwarf material would have a mass of about one tonne – equivilent to one million times the density of water.

If a star begins its life with a mass of more than about eight solar masses, however, its fate is far more violent. When the star has used up all of its nuclear fuel, even though it has lost some of its mass during old age, the weight of the remaining mass pressing down on the core makes it shrink into a ball even more compact than a white dwarf. It forms a neutron star, which contains about as much mass as our Sun in a sphere less than ten kilometres across. This collapse releases gravitational energy in the form of heat, just as the collapse of a gas cloud to form a star like the Sun releases heat (see page 32) – but very much more dramatic. The energy released blows the outer layers of the star apart in a spectacular explosion called a supernova. A supernova releases the same amount of energy as a million billion billion billion billion light bulbs. The remains of a supernova (a supernova remnant) unsurprisingly remain visible for thousands of years. Meanwhile, at the core of the supernova remnant the neutron star spins rapidly on its axis. A neutron star is a single ball of neutrons, and has the same density as the nucleus of an atom. It is in effect a single atomic nucleus several kilometres in diameter. The matter in a neutron star is in the most compact form that matter can exist. The density in a neutron star is one million times greater than in a white dwarf and one thousand billion times greater than water, so each cubic centimetre of neutron star matter has a mass of about one billion tonnes.

If matter is squeezed into an even more compact state than a neutron star – perhaps if a neutron star itself accretes enough material from its surroundings that it starts to collapse under its own weight – there is nothing to stop it collapsing completely to a point (called a singularity) and becoming a black hole.

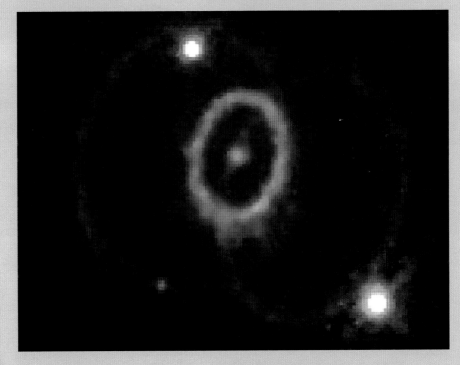

Supernova 1987A
In 1987 a supernova occurred in a satellite galaxy of the Milky Way. The shell of material blown off the star is in the middle of the picture. the two larger rings are thought to be caused by a pulsar created in the explosion.

material falling onto the neutron star and making it spin faster would have formed a swirling accretion disc around the star.

The three extrasolar planets must have formed in that disc in a way reminiscent of the way Mercury, Venus and Earth (and the rest of the planets of our Solar System) formed when the Sun was young. An accretion disc around a star with roughly the same mass as our Sun has produced three planets with similar masses and in similar orbits to the three inner planets of the Solar System. It is hard to regard this as merely a coincidence, and suggests that planets with roughly the same mass as the Earth and Venus are likely to be found in roughly similar orbits to those of the Earth and Venus. They would also be orbiting stars with masses roughly the same as that of our Sun – including other orange-yellow G-type stars.

As a postscript, while we were writing this book another pulsar, PSR B1620-26 has been found to have a planet as well. The information on the planet is rather sketchy at the moment, but it seems to have a mass of about 1.2 to 6.7 times the mass of Jupiter and an orbit between 10 and 64 au. More observations of this pulsar over a long period (as the orbital time of the planet is long) will pin down these numbers, but it looks as if pulsar planets are not an isolated phenomenon.

2.4 New worlds

While the pulsar planets were the first extrasolar planets to be discovered, they were not what we normally think of as true planets (i.e. they are not in orbit around stars like the Sun). However, for the discovery of a planet around a normal star we did not have to wait long.

Doppler planet searches had been ongoing since the late 1980s. Spectra taken by the telescopes were complex and required a lot of computational effort to analyse them. But two Swiss astronomers, Michel Mayor and Didier Queloz, working on a telescope in France had improved the technique. Instead of having to go away and analyse their data, the Doppler shift would appear on their computers in the observatory straight after the observations were complete. Mayor and Queloz chose 142 stars to observe all of which were similar to the Sun (spectral types G and K). In 1994 and 1995 they observed the star 51 Pegasi (hereafter

Lovell Telescope
The Lovell Telescope at Jodrell Bank was one of the first and most famous radio telescopes.

Astronomers sometimes make mistakes. In 1991, before the discovery of the planets around PSR B1257+12, Andrew Lyne and his colleagues at the University of Manchester thought they had detected a single planet orbiting another pulsar, PSR B1829-10, using the famous Jodrell Bank telescope. It soon turned out that the discovery was a mistake, which occurred because a slightly incorrect position for the pulsar had been entered into the computer used to search for the wobbles caused by planets. Instead of publishing a correction as quietly as possible, buried in the small print of a technical journal,

Lyne owned up to the mistake in a public lecture at one of the biggest astronomical events of the year, the meeting of the American Astronomical Society. Then a paper from the team appeared in the journal Nature, the world's most widely read science journal, under the title 'No planet orbiting PSR 1829-10'. This article included the confession 'we must accept full responsibility for this error'. The incident is worth mentioning as a shining example of the honest way in which science ought to be done – and also because after this the PSR B1257+12 data was double-checked for any similar mistakes and passed those tests with flying colours.

referred to as 51 Peg) and found a regular cycle of 60 metres per second in the star's velocity every 4.2 days. They had discovered the first true extrasolar planet (see box page 67).

It seemed clear that a true extrasolar planet had been discovered, the trouble was that it was a very strange planet. With a mass of at least 0.6 Jupiter masses (see footnote) it orbits 51 Peg at only 5 per cent of the Earth-Sun distance (0.05 au) every 4.2 days. The closeness of the planet to the star made the Doppler shift high (60 metres per second) and therefore relatively easy to find. Since the orbital period was so short it was easy to collect data from lots of orbits to be sure that it was a real signal due to a planet. However nobody had been expecting Jupiter-mass planets so close to their parent stars. It was so at odds with what people were expecting that many astronomers did not believe the results. Jupiter-mass planets form far from the star, they could not possibly form so close to the star. Look at our Solar System – the closest massive planet (Jupiter) is 5.2 au from the Sun. The planet of 51 Peg would be heated to about 1,000 degrees Celsius.

It did not take long for more planets to be discovered and then people had to take notice of the very unusual things these planets were telling us. Almost immediately Geoff Marcy and Paul Butler found another planet, 2.4 times the mass of Jupiter, around the star 41 Ursae Majoris at a distance of 2.1 au from the star.

Then, in December 1995 Marcy and Butler found a companion to 70 Virginis, which had a mass of 8.4 times the mass of Jupiter. This made it very big for a planet, almost brown dwarf sized. It was similar to a brown dwarf found around the star HD 114762 discovered in 1988 by David Latham, which had a mass of 11 times the mass of Jupiter. At the time many people thought that the inclination of HD 114762 was very low and so the actual mass would be a lot higher, classifying it as a brown dwarf.

What was also similar about these two large objects was the eccentricity of the orbits. Eccentricity is a way of measuring the departure from a circle of an orbit. The orbits of the planets in the Solar System are pretty circular (with the exception of Pluto and, to a smaller degree, Mars). The Earth's orbit has a small eccentricity which means that during summer in the northern hemisphere the Sun is actually further away than during

footnote
For the rest of this book we will give the minimum mass of planets calculated if the inclination were 90 degrees. The actual mass of the planets would generally be slightly higher than this but in most cases we do not know the inclination.

Extrasolar planet

This artist's impression of a giant extrasolar planet displays many of the features of Saturn. Many of the new planets that have been found to date are thought to be gas giants that lie relatively close to their star and are therefore often referred to as 'hot Jupiters'.

Planet discoverers

Michael Mayor and Didier Queloz, the discoverers of the first 'true' extrasolar planet around the star 51 Pegasi in 1995.

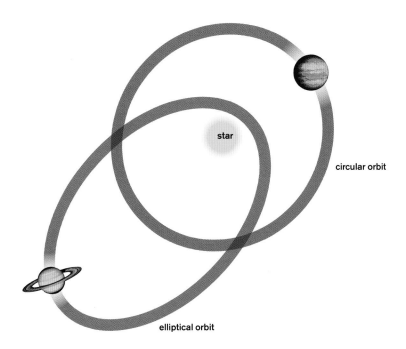

star

circular orbit

elliptical orbit

Circular and elliptical orbits
Here a Jupiter-like planet is seen in a circular orbit around a star, while the Saturn-like planet has a highly elliptical orbit.

winter (not that we notice this as the amount is negligible). The objects (planets or brown dwarfs) around HD 114762 and 70 Virginis both have very high eccentricities.

In fact, it was these high eccentricities that helped the planet hunters find them. When a planet is in a highly eccentric orbit, it pulls the star far more at closest approach (perihelion) than at furthest distance (aphelion) from the star. This extra gravitational tug at perihelion enhances the wobble, making it larger than if the planet were on a circular orbit. This is also how we know the orbit is eccentric – if the pull is more at one point in the orbit than at another this can be seen in the velocity curve.

The two objects around HD 114762 and 70 Virginis are now counted as massive planets rather than small brown dwarfs (so really David Latham in 1988 found the first extrasolar planet).

In 1996 Marcy and Butler announced three more planets, one each around the stars Rho Canceri A, Tau Bootis A and Upsilon Andromedae. These planets had masses of 1.0, 4.9 and 0.8 times that of Jupiter and they were in circular orbits close to their stars at only 0.11, 0.05 and

Michel Mayor and Didier Queloz were making their planet search using the Haute-Provence Observatory. In September 1994 they started to observe the star 51 Pegasi in the constellation of Pegasus. By January 1995 they noticed that 51 Peg showed wobbles in its velocity that could be caused by a planet orbiting once every four days.

The Swiss team waited until October, collecting data and spending a long time waiting while 51 Peg disappeared behind the Sun and could not be observed. When 51 Peg reappeared they found that the velocity variations were still there and occurred exactly where they had predicted. What they found was a regular cycle in the velocity of the star every 4.2 days of 60 metres per second.

However, there was a problem in that Mayor and Queloz did not know the inclination of the 51 Peg system. It is difficult to measure the mass of a planet discovered from the Doppler shift without any other information, as what is being measured is usually only part of the wobble induced by a planet. The velocity measured is the velocity in our line-of-sight. Any part of the velocity not in our line-of-sight is unknown.

If Mayor and Queloz's planet was at a high inclination, viewed almost face-on, then the 60 metres per second velocity shift they observed would only be a tiny part of the total shift and the mass of the planet would be greater than ten times the mass of Jupiter. This would classify it as a brown dwarf. However, they calculated that the likelihood of the system being inclined so little as to make the mass of the planet more than four times the mass of Jupiter was less than 1 per cent.

Their argument that they were not looking at a face-on system was aided by observations of the star. If we were to look at a star side-on the spectral lines would be smeared out slightly. This is because the light from the side moving towards us as the star rotates is red-shifted, while light from the side moving away from us is blue-shifted. If the star is viewed face-on, as it would be in a low inclination system, this smearing would not happen. The spectral lines of 51 Peg showed some smearing indicating that Mayor and Queloz were indeed looking a the system side-on (exactly what the inclination was they did not know, but it had to be quite high). If the inclination was 90 degrees the mass had to be 0.6 times the mass of Jupiter, and as the inclination increases this mass would increase slightly.

In October 1995, Mayor and Queloz announced that they had found a true extrasolar planet. One of the teams that had been looking for planets for several years, headed by Geoff Marcy and Paul Butler in the US (more about these two later), went and observed 51 Peg immediately. They found exactly the same wobble that Mayor and Queloz had found.

Mayor and Queloz submitted their findings to the scientific journal Nature. Nature's policy is that authors cannot talk about their findings or release information to the press before the paper is published. This led to a misunderstanding as the press interviewed Marcy and Butler about the discovery and they (through no fault of their own) were often wrongly credited with the discovery, especially by the US media.

The media circus that followed this announcement was understandable; at last we had proof that a planet circled at least one other star.

Planetary signature
As time passes (phase) the velocity of a star that is orbited by a planet changes rhythmically. When the velocity is plotted, the undulating curve on the graph signifies the presence of a planet, as seen in the velocity curve of 51 Peg.

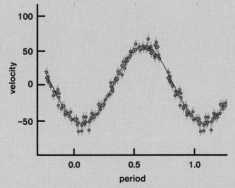

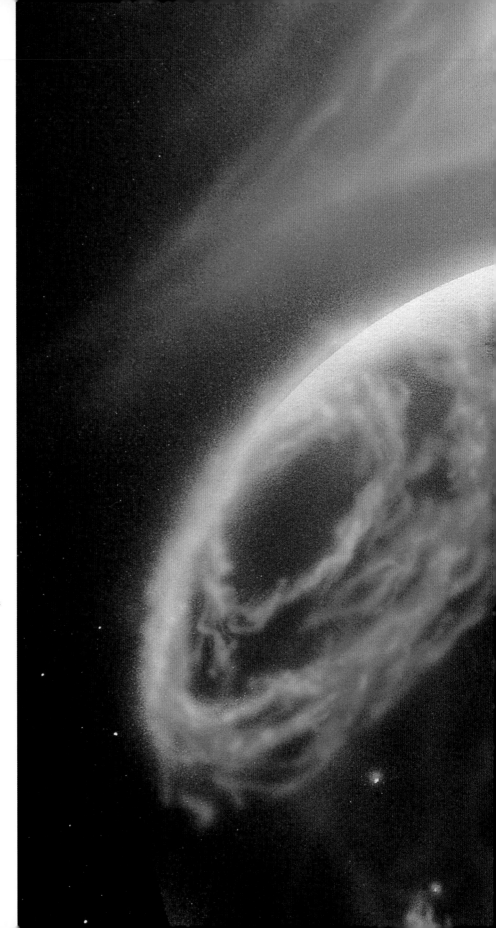

The 51 Peg planet

An artist's impression of a
planet around 51 Peg.
The star is huge and the
planet, which is about the
size of Jupiter, is situated
very close to it. An auroa
can be seen at the planet's
pole and in the foreground
is an asteroid.

Extrasolar planets

This is a list of all the extrasolar planets found by the end of the year 2000. With new planets being discovered all the time this list is bound to be out-of-date by the time you read this, but it gives a good idea of the sorts of planets that have been found.

For each star the distance from us in light years is given as well as the spectral class of the star.

For each planet around each star (normally just the one) the minimum mass the planet can have is given (this depends on the inclination) in units of the mass of Jupiter, except for the planets around PSR 1257+12 which are small enough to have masses measured in units of the Earth's mass.

The length of the planets orbit is then given, normally in days, but sometimes in years.

Finally the eccentricity of the orbit, the larger this number the greater the divergence from a circle.

Where there are '?'s we are unsure of the values.

Name	d/ly	type	mass	r/AU	p/d	ecc.	
HD 83443	141	K	0.35	0.04	3.0	0.08	
			0.16	0.17	30.	0.42	
HD 16141	117	G	0.22	0.35	76.	0.28	
HD 168746	140	G	0.24	0.07	6.4	0.00	
HD 46375	109	K	0.25	0.04	3.0	0.00	**5**
HD 108147	126	G	0.34	0.10	11.	0.56	
HD 75289	94	G	0.42	0.05	3.5	0.06	
51 Peg	50	G	0.47	0.05	4.2	0.00	
BD-10 3166	?	G	0.48	0.05	3.5	0.00	
HD 6434	131	G	0.48	0.15	22.	0.30	**10**
HD 187123	163	G	0.52	0.04	3.1	0.03	
HD 209458	153	G	0.69	0.05	3.5	0.00	
ups And	44	F	0.71	0.06	4.6	0.03	
			2.11	0.83	241	0.18	
			4.61	2.50	3.5y	0.41	**15**
HD 192263	65	K	0.76	0.15	24.	0.03	
eps Eridan	10	K	0.86	3.30	6.8y	0.61	
HD38529	137	G	0.81	0.13	14.	0.28	
HD 179949	88	F	0.84	0.05	3.1	0.05	
55 Cnc	41	G	0.84	0.11	15.	0.05	**20**
			>5	>4	>8yr	?	
HD 121504	145	G	0.89	0.32	65.	0.13	
HD 37124	108	G	1.04	0.59	155	0.19	
HD 130322	98	K	1.08	0.09	11.	0.05	
rho CrB	57	G	1.1	0.23	40.	0.03	**25**
HD 52265	91	G	1.13	0.49	119	0.29	
HD 177830	192	K	1.28	1.00	391	0.43	
HD 217107	64	G	1.28	0.07	7.1	0.14	
HD 210277	69	G	1.28	1.1	437	0.45	
HD 27442	59	K	1.43	1.18	440	0.02	**30**
16 Cyg B	70	G	1.5	1.70	2.2y	0.67	
HD 134987	82	G	1.58	0.78	260	0.25	
HD 160691	50	G	1.97	1.65	2.0y	0.62	
HD 19994	73	F	2.0	1.3	454	0.2	
Gilese 876	15	M	2.1	0.21	61.	0.27	**35**
HD 92788	105	G	3.8	0.94	340	0.36	
HD 82943	90	G	2.24	1.16	443	0.61	
HR810	51	G	2.26	0.93	320	0.16	
47 Uma	46	G	2.41	2.10	3.0y	0.10	
HD 12661	121	K	2.83	0.79	265	0.33	**40**
HD 169830	118	G	2.96	0.82	230	0.34	
14 Her	59	K	3.3	2.5	4.4y	0.35	
GJ 3021	57	G	3.31	0.49	134	0.51	
HD 195019	122	G	3.43	0.14	18.	0.05	
Gilese 86	36	K	4.23	0.11	16.	0.05	**45**
tau Boo	51	F	3.87	0.05	3.3	0.02	
HD 190228	202	G	4.99	2.31	3.1y	0.43	
HD 168443	287	G	5.04	0.28	58	0.54	
			15?	2?	4.5y	0.28	
HD 222582	137	G	5.4	1.35	1.6y	0.71	**50**
HD 10697	98	G	6.59	2.0	3.0y	0.12	
70 Vir	59	G	6.6	0.43	117	0.4	
HD 89744	130	F	7.2	0.88	256	0.7	
HD 114762	132	F	11.	0.3	84.	0.33	**54**
BROWN DWARFS							
HD 162020	53	K	13.7	0.07	8.4	0.28	
HD 110833	55	K	17	0.8	270	0.69	
BD-04 782	?	K	21	0.7	241	0.28	
HD 112758	54	K	35	0.35	103	0.16	
HD 98230	?	F	37	0.06	4.0	0.00	
HD 18445	?	K	39	0.9	1.7y	0.54	
HD 29587	147	G	40	2.5	3.2y	0.00	
HD 140913	?	G	46	0.54	148	0.61	
HD 283750	54	K	50	0.04	1.8	0.02	
HD 89707	82	G	54	?	198	0.95	
HD 217580	59	K	60	1?	455	0.52	
Gilese 229	22	M	40	40?	>200y	?	
PULSAR PLANETS							
PSR 1257+12	1000	–	3.4e	0.36	66.5	0.02	
			2.8e	0.47	98.2	0.03	
			100e	40?	170y?	?	
PSR B1620-26	12000	–	1.2–6.7	10–64	62–389y	?	

CM Draconis is a binary star system about 57 light years away. It was chosen as a target for a transit search because the two stars in the system are eclipsing binaries: they pass in front of one another as they orbit. This means that the plane of the orbits is very close to edge-on so any planets around the stars will transit – hopefully making them visible.

In 1995 CM Draconis was targeted and has been observed regularly ever since to search for a dimming of the light from the stars. In 1996 a possible Jupiter-sized planet was found causing an 8% dimming of the light. This planet is not fully confirmed as definite but the chances of its existence are high.

More excitingly a number of dimming events were found that could correspond to Earth-like planets orbiting close to the central binary stars. More observations led to most of these candidates being ruled out as they did not re-occur (as you would expect from a planet orbiting the stars). There does remain the possibility that two Earth-like planets, with diameters about twice Earth's, are orbiting CM Draconis. The planets would be far closer to the stars than Earth is to the Sun, orbiting every 40 to 60 days.

The researchers themselves are rather sceptical about their results at the moment, unlike the somewhat overblown media coverage that suggested that the planets were definitely there.

CM Draconis
The white blob right of centre is the binary star CM Draconis. Other stars nearby are also observed and their brightness is compared to that of CM Draconis to see how it changes. This gives a far more accurate estimate of its brightness from the Earth's surface, which often varies as the weather changes – not good when we are looking for tiny variations.

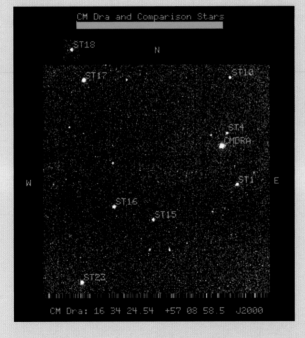

0.06 AU respectively. Upsilon Andromedae proved to have further surprises when it was found to have not one, but three, giant planets in orbit around it (see box page 73).

At the time of writing about 60 planets have been found around other stars. New planets are being discovered all the time; in early August 2000, ten new planets were announced. We are sure that by the time you are reading this book this number will be out of date. However, we have brought together all the current extrasolar planets in a table, at the time of writing.

The number of planets is now big enough for us to say quite a lot about planets around nearby stars. But, as with many new discoveries in science, the extrasolar planets have raised more questions than they have answered.

There seem to be two distinct classes of planets – the hot Jupiters, and highly elliptical planets like 70 Virginis and HD 114762. The first type are on circular orbits very close to the star, while the second are on highly elliptical orbits further away from the star.

About 800 nearby stars have been studied whilst looking for planets using both Doppler methods and astrometry. Most of these stars are similar to the Sun in spectral type and about one in twenty have a planet. One in particular; Upsilon Andromedae has three!

There is one type of system that is lacking from this ever growing list of planets so far – planetary systems that look like our own, with a massive planet (or four) in circular orbits far from the star. Does this mean that our Solar System is very unusual?

Possibly not, as we have said, planets close to the star and planets on elliptical orbits are the easiest to find. They create a far bigger wobble than do massive planets on distant circular orbits. The new technologies being used to hunt for planets have a high enough accuracy to find systems like our own if they exist, but to have good detection the star must be observed for at least one orbital period, preferably several. The orbital period of Jupiter is eleven years, therefore if there are planetary systems like our own we should find them, but we may have to wait a few years yet.

This is an example of a complication called a selection effect that is always a problem in astronomy. We have to be careful what conclusions

The Upsilon Andromedae system was one of the first stars to be observed with an extrasolar planet. The velocity of the star showed variations due to a planet of about 0.7 times the mass of Jupiter orbiting 0.06 au from the star. However, this did not fully account for the velocity variations that were observed so the curve due to this planet was removed – and another planet appeared, this time with a mass of 2.1 Jupiter's at a distance of 0.8 au from the star. When the process was repeated yet another planet appeared with a mass of 4.6 Jupiter's at a distance of 2.5 au, this time in a slightly elliptical orbit. Since then HD 83443 has shown two massive planets and 55 Cnc has a possible double planet system.

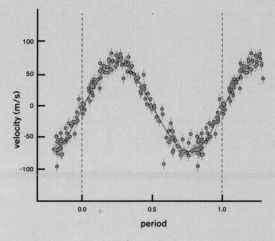

The velocity curves of Upsilon Andromadae

The top graph shows the raw velocity curve of Upsilon Andromedae.

In the second graph, the velocity of the closest planet has been removed and another planet appears.

In the third graph the velocity of the the second planet has been removed, and the resulting curve shows that yet another planet is present.

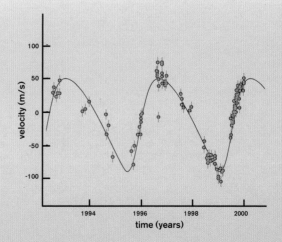

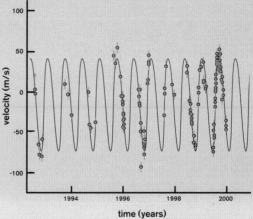

we draw from the information we have on planets because that information is not complete. One thing we can say for sure is that massive planets are not rare. At least one in twenty stars have planetary systems, and some of those that have not shown a planet yet may do so in a few years time.

2.5 Making planets

The planets of our Solar System formed from a disc of material swirling around the young Sun, while the Sun itself was still shrinking and achieving the stable state that it is in today. As the Sun formed from a collapsing cloud of gas and dust in space, it would have spun faster on its axis. An object of a certain mass has a certain size and has a certain amount of a property called angular momentum. Angular momentum is always conserved, so if the object shrinks it will spin faster. The spin creates a centrifugal force, which tends to stop the object collapsing. The Sun could never have shrunk from a cloud of interstellar gas into a star if it had not been able to dispose of some of its angular momentum. One of the ways it achieved this was by losing mass, with material streaming off into space, carrying angular momentum with it. Another way was by transferring angular momentum from the proto-Sun to a disc of material surrounding it, a disc in which the planets would form. The fact that the planets all orbit the Sun in the same direction, and the Sun itself rotates (every 25.3 days) in that direction, is clear evidence for this.

The Orion Nebula
Located 1,500 light years away, the Orion Nebula is forming thousands of Sun-like stars. Around some of these stars discs have been found. In these discs planetary systems may well be forming.

When we say that the Solar System formed from a cloud of gas and dust in space, this may give a misleading impression of the huge importance of the dust. Interstellar material consists of 98 per cent gas (almost all of that hydrogen and helium), and only about 2 per cent everything else. The dust is formed from these heavy atoms that have been manufactured inside stars through the process of nuclear fusion, and scattered through space when those stars died – it is, literally, stardust. The hydrogen and helium are remnants from the Big Bang in which the Universe was born. When the cloud that would eventually begin to form the Solar System started to collapse, the particles of dust in the cloud were as fine as the solid particles in cigarette smoke. However, even before some of this material had begun to settle into a disc around the young Sun, these tiny grains began to collide with one another, more regularly as the density of the collapsing cloud increased. This enabled the grains to stick together to make fluffy supergrains, each a few millimetres in size.

You might think that such delicate grains (a bit like snowflakes) would have been broken apart as the density of the cloud increased, and the collisions became more violent. But because the dust settled into a disc around the Sun, moving in the same direction as the Sun's rotation, there were very few head-on collisions. Intead there were many gentle bumps when one grain overtook another. As the grains formed bigger and bigger masses, their own gravity began to be important, pulling other grains towards them and sweeping up material from the disc to form planetesimals, rather like the objects in the asteroid belt.

While all this was going on, the Sun was increasing in temperature at the core of the disc. Close to the Sun, the heat was strong enough to drive out gas and melt grains of icy material. Hence the planetesimals that eventually came together to form planets were composed chiefly of

Simulated star birth
This computer simulation of star formation shows discs forming around young stars. These simulations were run by the Cardiff Star Formation Group, of which Simon Goodwin is a member. The simulations are not yet good enough to see planets forming in the discs, but we are reasonably sure they would. Simulations of small sections of discs form many planets and moons.

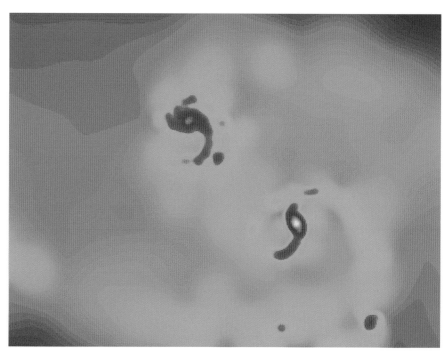

elements such as iron and compounds such as silica (which are not easily vaporised), forming the four inner planets. Further away from the Sun, it was cool enough for icy grains to stay frozen, so that the giant planets are richly endowed with compounds such as methane and ammonia. All the while, the heat from the central solar source was driving the 98 per cent of the original disc material, in the form of gas, out into space, carrying angular momentum away with it.

Computer simulations suggest that, in the final stages of the planet-building process (just before the four inner planets took on the form we know today), there were more than one hundred objects the mass of the Moon in the inner part of the Solar System. In addition, there were at least ten with masses exceeding that of Mercury, and five to ten with masses similar to Mars. A series of catastrophic collisions between these objects formed Venus and the Earth over a period of about ten to fifty million years, more than four billion years ago. In one of the last of these great collisions, an object at least the size of Mars struck the Earth a glancing blow. This melted both the incoming object and the entire surface of the Earth and splashed material out into space, where some of it formed a ring around the Earth. Our Moon formed from the material in this ring, coalescing in a manner reminiscent of the way the planets themselves had formed from a disc of material in orbit around the Sun.

Protoplanetary discs This explanation of how the planets of our own Solar System formed fits in neatly with the discovery of dusty discs of material around more than one hundred other young G- or K-type stars. The first such discs were only detected in 1984, when the Infrared Astronomy Satellite (IRAS) went into orbit around the Earth and was used to study such stars at infrared wavelengths (amongst other things). Infrared radiation is blocked by the Earth's atmosphere so it can only be studied from space. It is radiated by relatively cool objects in the universe such as clouds of gas, and indeed, planets themselves. However, the discovery that extensive discs of dusty material surround many young main-sequence stars came as a complete surprise to astronomers. The discs only radiate energy at all because the heat of their central stars is warming them up. Once the discs had been detected at infrared wavelengths, the Hubble Space Telescope, and subsequently ground-based telescopes, were used to photograph the discs using visible light. This is very tricky, because the light from the central star tends to blot out the faint image of the disc – it is like trying to photograph a pocket torch alongside a searchlight. The starlight has to be subtracted from the image electronically in order to reveal the disc. The first such disc to be imaged in this way orbits a star known as Beta Pictoris, which is about 53 light years away from Earth. The disc is about 1,000 au across, but it has a hole in the middle, which seems to have been swept clear of dust. The hole extends for about 15 au on either side of the star, roughly the size of the region of our Solar System occupied by the planets. This strongly suggests that the hole has been swept clear of dust by planets forming around Beta Pictoris.

The total mass of the dust in the disc around Beta Pictoris is about the same as the mass of Jupiter, although individual dust grains can only be a few millionths of a metre (a few microns) across, judging from the spectroscopic evidence. There is also little gas in the disc, equivalent to

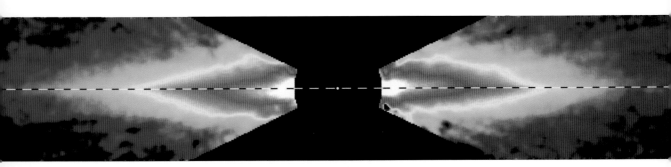

**The gas rings of
Beta Pictoris**
The disc of gas and dust around Beta Pictoris was the first such feature to be found. In this picture, the star at the centre has been blanked out to allow the disc to be observed.

about one-hundredth of the mass of the Earth. Clearly, most of the primordial gas has been blown away into space.

In total, about half of the stars with ages less than about 300 million years that have so far been studied in this way have been found to possess dusty discs similar in size to our Solar System. One of the best examples of such a disc is provided by the young star HL Tauri, which has a disc containing about one-tenth as much mass as our Sun (about ten times the total mass of all the planets in the Solar System), extending over a diameter of 2,000 au. This is a very young system (much younger than even Beta Pictoris), and the rotation of the disc can actually be measured using the Doppler technique.

Another large rotating disc of dusty material, more than ten times the diameter of the orbit of Neptune, has been found around a star known as MWC480, about 450 light years away from Earth in the direction of the constellation Auriga. MWC480 is both brighter and more massive than our Sun. This disc was first detected at radio wavelengths, is predicted to evolve into a system like the Beta Pictoris disc, which is a few tens of millions of years old. The fact that these discs do evolve, and eventually lose most of their material to interstellar space, has been confirmed by studies using the Hubble Space Telescope and the Infrared Space Observatory (a successor to IRAS). These studies showed that about 60 per cent of individual stars younger than 400 million years have dusty discs around them, but only 9 per cent of stars older than 400 million years have such discs. It seems that 90 per cent of stars that start out with dusty discs lose them after about 300–400 million years. This closely

Discs in Orion Nebula
Over 300 protoplanetary discs like the two shown here have been found around Sun-like stars in the Orion Nebula in which there is a good chance that planets could form.

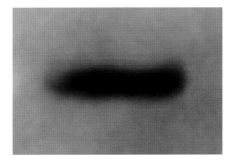

matches the evidence of how and when planets formed in our Solar System (within a few hundred million years of the Sun 'switching on'), which is inferred from studies of rocks and meteoritic material. The size of the disc in stars like Beta Pictoris also compares with the size of the Kuiper Belt in our own Solar System (see page 23), although the Kuiper Belt contains only about one-hundredth as much mass, suggesting that it is the remnant of a Beta Pictoris-type disc. Putting this information together, it seems that our Solar System is far from being unusual, but is a typical product of the way stars like the Sun form. As a conservative estimate, from the number of dusty discs we can see, we should expect at least half of all the stars like our Sun to possess families of planets similar to our Solar System. All we have to do is find them!

Problem planets What can the planets we have already found tell us about the planet-forming process? An interesting feature that has been observed in the types of stars that have planets so far is that these stars seem to be abnormally rich in heavy elements. Approximately 2 per cent of the Sun consists of elements other than hydrogen and helium and likewise, the stars that have planets are all found to have similar, or higher proportions of heavy elements. Could this be telling us that planet formation requires lots of heavy elements?

Obviously small rocky planets like the Earth contain elements that are heavier than hydrogen and helium, but massive gas planets like Jupiter are mainly hydrogen and helium. Do Jupiter-type planets need rocky cores to build around?

This also means that planet-bearing stars are generally quite young. Heavy elements are added to interstellar gas by the deaths of stars, so with time the amount of heavy elements in the gas between stars increases. This gas congregates in clouds, which then collapse to form new stars, which are richer in heavy elements than the previous generation. Therefore the amount of heavy elements a star has is a very crude cosmological clock, giving a rough idea of its age.

The existence of hot Jupiters has made astronomers rethink our understanding of planet formation. As we have seen, we believe that the Solar System formed from a disc of gas and dust surrounding the young Sun. In this standard picture small rocky planets form close to the star and massive gas giant planets like Jupiter form several au away from the star, where it is cooler. This picture certainly fits our own Solar System but is very much at odds with many of the extrasolar planets.

Before the discovery of hot Jupiters it was not believed that gas giants could exist that close to the star. This was because the main components of massive planets are hydrogen and helium and the heat from the central star should be enough to evaporate such light elements at a close distance from the star, heating them enough so that they escape. Only far from the star could such elements remain to collect and form gas giants like Jupiter. Our own Solar System was the basis for this view – the gas giants form a long way from the Sun.

So how do massive planets get so close to the star? There are two possibilities: the planets form close to the star, or they form further out and then move closer towards the star. Even before hot Jupiters were discovered, astronomer Douglas Lin had suggested that massive planets could form far from their central star but migrate towards the star. Planets drag against the other material in the proto-planetary disc and lose energy. As a result they lose speed and move closer towards the star.

One problem with this suggestion is that it is not clear why massive planets should move to distances close to the parent star and not be pulled into it. Indeed, some astronomers have suggested that it is the cannibalism of planets that gives the stars with planets their high heavy element abundance. The heavy elements in stars tend to be pulled towards the centre of the star as they are attracted by gravity more than light elements. The infall of planets into a star would replenish the atmosphere of the star with heavy elements.

Another problem may be making planetary systems like our Solar System. In some models massive planets always move closer towards the star, never staying where they formed.

The picture is far from clear. Nobody has found a method of forming massive gas giant planets close to a star, so the migration picture is the best we have. Then again, there is no satisfactory model of planet formation, so maybe there is a way of producing massive planets close to a star. The new information provided by the discovery of extrasolar planets has changed the way astronomers think about planet formation, but the subject is now so revolutionized that it will probably take many years to digest fully.

Binary star planets Another very odd feature that has been observed is that binary stars can have planets. We mentioned earlier (Section 2.1) that systems of two or more stars orbiting each other were thought to be unlikely homes for planetary systems. The combined attraction of the two

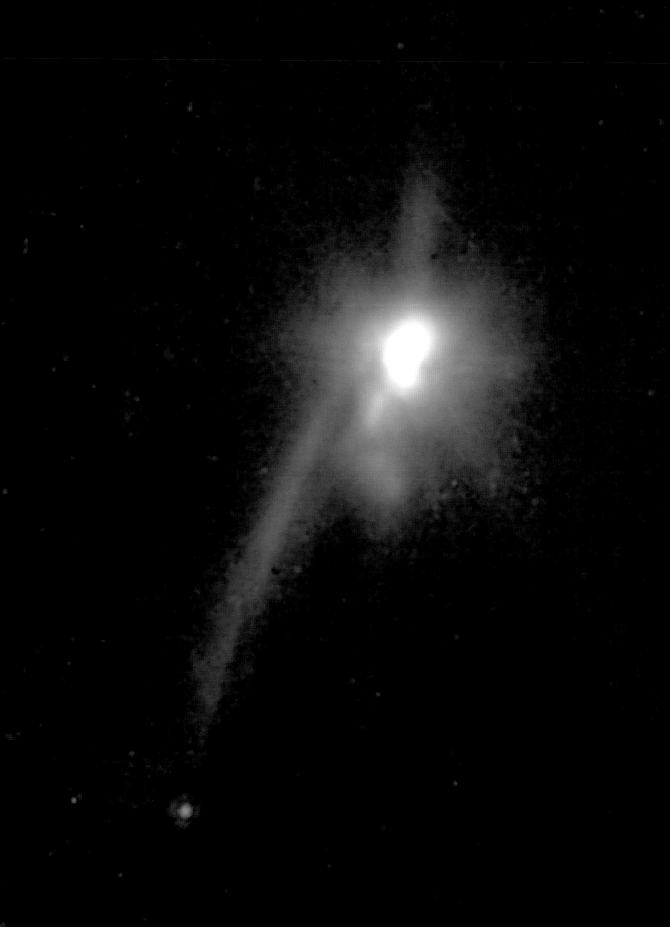

stars would eject any planets from the system. This appears to have happened in TMR-1, the two bright objects at the centre are a young binary system but the faint object at the bottom could well be a planet ejected from the system, only visible because it is now 1,400 au from the stars. This picture could well be the first image of an extrasolar planet. Unfortunately we do not know the mass of the object and so cannot be sure; it may be a brown dwarf.

However two of the very first stars to be found to have planets were binary stars: Rho Canceri A and Tau Bootis A. Both of these stars have faint red dwarf companions orbiting far from the main star. It appears that distant binary companions do not disrupt the planet-forming process. Even stranger is a more recent planet found around an unnamed star, which appears to orbit both stars in a binary system. This does not fit at all well with our current models of planet formation.

Extrasolar planet?
This image, captured by the near infrared camera of the Hubble Space Telescope in 1997, may contain the first actual image of an extrasolar planet. The faint blob in the bottom left of the picture has been ejected by the bright binary system above it, and is considered too faint to be a red dwarf.

Hungry star
This artists impression
shows a star swallowing
a Jupiter-like planet. If
planets do migrate closer
to a star many might
not stop to become hot
Jupiters, but would
instead be consumed
by the star. Some
astronomers think that
planet eating may
account for the high
heavy element content
of planet-bearing stars.

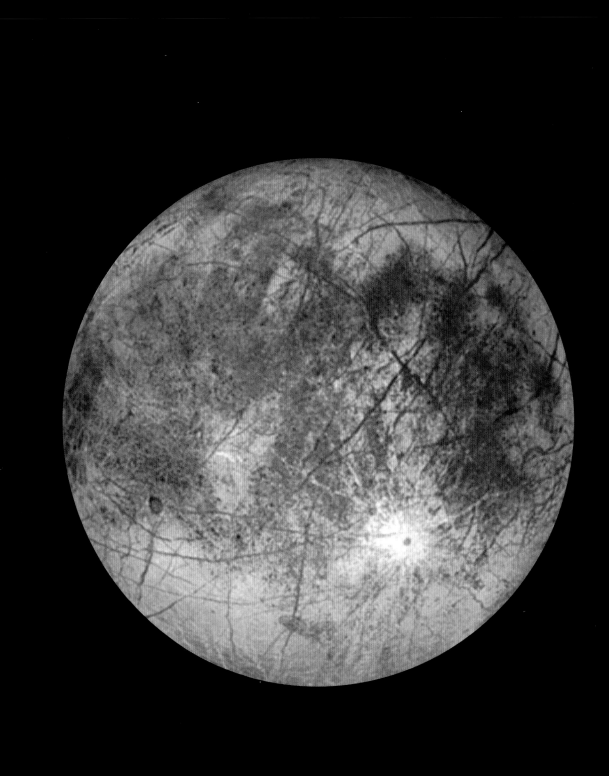

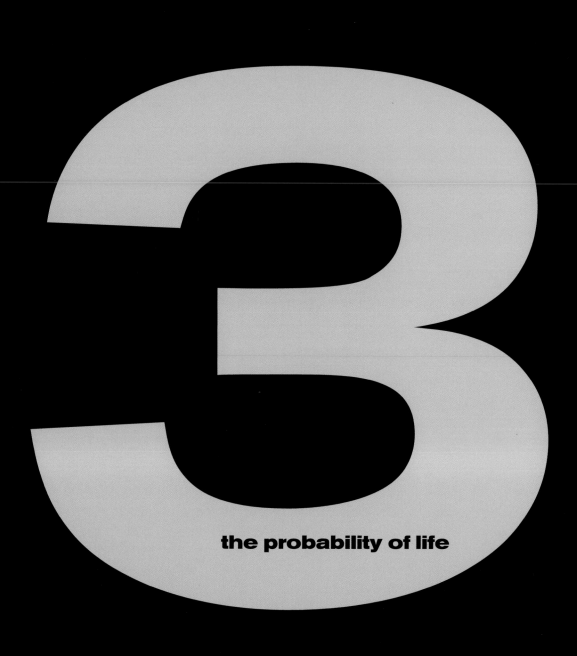

3

the probability of life

3.1 Future planet searches

With the exception of the pulsar planets, all of the planets discovered so far have been massive Jupiter-like planets. For the proverbial 'person on the street' the word planet does not just mean the astronomical definition of an object less massive than ten times the mass of Jupiter but a world like the Earth with which we are all (hopefully) so familiar. As was described in Section 1.1, planets like the Earth are known as terrestrial planets; small, rocky planets a few thousand kilometres across. In our own Solar System, half of the planets (or more than half if we include Pluto as a planet) are terrestrial planets. Once we include the large moons of Jupiter and Saturn as well, the Solar System contains many large rocky worlds.

The predominant reason for the interest in terrestrial planets compared to gas giants, for both astronomers and the general public alike, is that terrestrial planets are the most likely to harbour life. As amazing as the discovery of extrasolar giant planets has been, what we really want to find are planets like the Earth.

Using Doppler and astrometric techniques we are pushing the boundaries of technology to discover giant planets. The best Doppler techniques can detect variations of a few metres per second, and astrometry can find variations of a few milliarcseconds. This would allow the detection of planets of a similar mass as Saturn. However, Saturn is still one hundred times more massive than the Earth.

While Jupiter shifts the Sun by 800,000 kilometres and gives it a velocity shift of 12.5 metres per second, the Earth gives the Sun a mere 450 kilometre shift and only a 1 metre per second velocity shift. The difference between the distance and velocity shift due to the Earth and Jupiter is due to the Earth being 5.2 times closer to the Sun and only taking one year for a single orbit, despite being 315 times smaller than Jupiter. As Jupiter moves the Sun at the speed of a sprinter, the Earth shifts it at a mere slow walk.

This tiny shift in position and velocity is impossible to find using current technology. Finding the massive gas giants that we have already found has pushed technology to the limit. All is not lost, however, as you

Gravitational lens
The massive cluster of galaxies in the foreground of this picture has lensed distant galaxies behind it, creating smered images that show up as arcs around the clusters. This magnifies these normally invisible galaxies so that we can see them.

will probably have guessed. Several near-future technologies and space missions should be able to locate terrestrial planets if they exist.

Microlensing and GEST One of the predictions of Einstein's General Theory of Relativity is that light should be bent when it passes close to a massive object (see box page 91). These gravitational lenses can be very strong, massively distorting background objects. However, the sort of lensing we are interested in is a lot less dramatic and is known as microlensing.

When an object (normally another star) passes in front of a distant star it causes an increase in the brightness of the distant star. It does this by focusing more of the light than normal towards us. Astronomers now monitor millions of distant stars to try and detect gravitational lensing, because it can tell us about the object passing in front of the star even if we cannot see it.

This monitoring can also detect planets. Gravitational lensing occurs when the 'lenser' (whatever it might be) is closer to the distant star than a critical distance known as the Einstein radius. This defines a circle around the lenser, known as the Einstein Ring. The light from the star changes in a very systematic way. If a planet is present inside the Einstein Ring then its light can also be amplified. This only happens very briefly but a sharp spike of extra light appears.

A gravitational lensing event called MACHO-98-BLG-35 was observed in 1998 as part of an ongoing experiment to look for brown dwarfs. When a star is lensed, the amount of light amplification and the timescale over which it occurs depends on the mass and speed of the lenser. With lots of events the number and distribution of brown dwarfs that are very difficult (often impossible) to observe normally can be determined. The interesting observation about MACHO-98-BLG-35 was the occurrence of a spike, which indicated the presence of a planet within the Einstein Ring.

The mass of the planet is probably between that of Earth and Neptune (the latter is seventeen times the mass of the Earth). Unfortunately this is the best possible estimate, but it does show that searching using microlensing can detect planets. The advantage of microlensing is that distance is not really a factor, as lensing can detect planets throughout the galaxy. Its main

A lensed planet?
A close-up of the magnification by a gravitational lens of what might be a planet. The graph shows how much the light from the star was amplified.

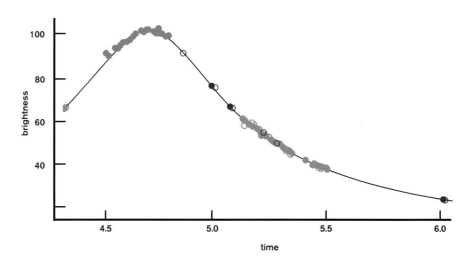

disadvantage is that it is a one-shot method. We can only identify the planets when a lenser passes close enough to the star, and the planet is in the right place. The chance of the same star being lensed again is virtually zero so we have no opportunity to double-check the first observation.

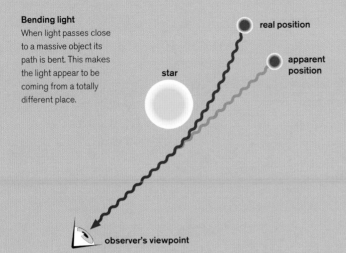

Bending light
When light passes close to a massive object its path is bent. This makes the light appear to be coming from a totally different place.

real position

apparent position

star

observer's viewpoint

Just as objects can have their path through space changed by the gravitational pull of another massive body, so can light. The General Theory of Relativity described this phenomenon, when it was published in 1915. The first prediction of relativity was tested in 1919 during a total solar eclipse. During such an event the Moon covers the Sun allowing stars close by to be seen, when normally their light would be swamped by that of the Sun. Stars very close to the Sun in the sky would have their light bent slightly, so that they appear to be in a different position to where they would be expected to be. This was exactly what occurred, and the size of their shift was exactly what was predicted by relativity.

Despite this drawback a satellite called the Galactic Exoplanet Survey Telescope (GEST) is being planned for launch in 2005 to search for planets. As with some of the other missions we will discuss, funding is not yet secured so it may never be launched or the launch date could be moved backwards.

The primary advantage of looking for planets with an instrument like GEST is that it would supply us with good statistics for the total number of extrasolar planets in the Galaxy. We have a good idea of the probability that if a planet is present it would reveal itself in a gravitational lensing event (the chance is about 10 per cent for a Jupiter-like planet at 5 au). Therefore if one in ten events show a Jupiter-like planet it would be likely that virtually all stars have such a planet. If only one in one million show a Jupiter-like planet then they are very rare. The chance of seeing an Earth-like planet is far smaller, but the same sort of argument follows.

Transit detectors Another way to look for planets is to look at their transits across the face of the star. When a planet moves across a star the brightness of the star drops slightly as a tiny fraction of its light is being obscured. What would be needed is a very sensitive optical telescope. As we are searching for changes in brightness it does not need excellent resolution, just an ability to detect small variations in the amount of light from a star.

The change in a star's light can be quite large for transits by massive planets that are close to the star (such as hot Jupiters) or infinitesimal if the planet is small. Jupiter would cause a 1 per cent reduction of light from the Sun lasting about thirty hours, detectable with current telescopes, but the Earth would only produce a 0.008 per cent dimming over about thirteen hours. The dimming effect of Mercury would be seven times worse than this.

The amount of light reduction would give an indication of the radius of the planet, while the duration of the dimming would give information about its orbit. Knowing the orbit and the mass of the central star, the mass of the planet can be determined. Since the planet is being observed in transit we know that the inclination of the system must be almost exactly side-on, meaning that the normal uncertainty in the calculation of the mass of the planet would not be present.

Unfortunately, only about 1 per cent of planetary systems will be side-on to us; the vast majority will have inclinations too great for this method to work. Another problem is that the telescope would have to be looking at the star at just the right time. Transits only last for a few hours, and in the case of the Earth happen once a year or for Jupiter only once every eleven years. To make matters worse, many stars naturally vary their brightness due to being far less stable than stars like the Sun, meaning that a small change caused by a planet may never be seen.

Despite these problems, the plan is to observe 100,000 stars so that up to 1,000 of them will show transits. If the planets are close in to the star then each year there will be a 2.5 per cent chance of seeing a transit in each star (if there is one to be seen). If close-in planets are common then this method should confirm this.

Three missions are planned. The European Space Agency (ESA) has approved plans for COROT (COnvection, ROtation and planetary Transits) to be launched in 2004. nasa has plans on the drawing board

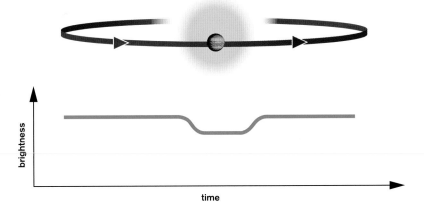

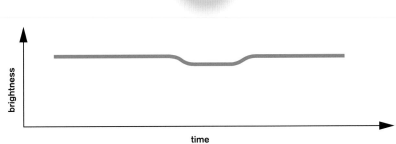

Planet in transit

When a massive planet moves across the face of a star, the light from the star drops for a few hours (top).

If the planet is small then the drop in light is smaller (middle).

However, if the star is active, the change in brightness will be impossible to see (bottom).

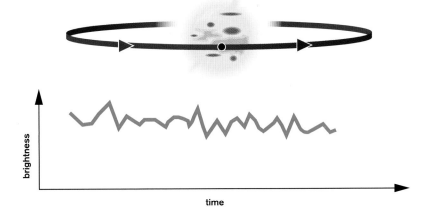

for a similar mission called Kepler (after the famous astronomer Johannes Kepler who discovered that planets orbit in ellipses, not perfect circles), which is hoped to be launched in 2004 as well. ESA also have another satellite, Eddington (after the famous British astronomer), which, like Kepler is not fully funded yet.

These three missions will do the same thing because these careful observations of stars will not just enable detection of planets, but they will also add considerably to our knowledge of stars. Most of the missions we mention in this section are not just planet detectors (although that is what we are concentrating on), but will also enable many other astrophysical observations.

Space Interferometry Mission (SIM) A satellite that will fly is the NASA Space Interferometry Mission (SIM). SIM uses a method called interferometry to turn several small telescopes into one large telescope. We'll be looking at interferometers in more detail later, for now how they work does not matter. The upshot is that SIM will be able to determine the positions of stars to incredible accuracy. It will be able to pin point the location of a star to a positional accuracy measured in thousanths of an arcsecond! Going back to our distance of a penny analogy from earlier, this is equivalent to the diameter of a penny at 2,000 kilometers.

SIM will take an astrometric approach to finding stars, looking for the tell-tale wobbles in a star's motion that a planet produces. Unlike current astrometry, SIM will be able to spot the tiny wobbles caused by terrestrial planets and determine their masses and orbits. In the near future, SIM is certainly the most exciting planet finding mission. SIM should find most Jupiter-mass and larger planets that are located within about 3,000 light years of our Solar System and all Earth-mass planets around the nearest 200 or so Sun-like stars.

Hopefully SIM will be launched in 2004, although launch times have a tendency to be put back.

GAIA (2009) In 2009, ESA hope to launch their GAIA mission. Like SIM, GAIA is an astrometric satellite but one that will identify the positions of the billion brightest objects in the sky. GAIA is more of a

◄ **SIM**
An artist's impression of one of the space telescopes that will be used in NASA's Space Interferometry Mission (SIM).

Some astronomers are hoping to see giant planets directly. As has been said, the light reflected from a Jupiter-like planet is far less than from the central star, but it is just possible to see in a special set of circumstances.

If the central star is very small and faint, a red M-type dwarf for example, it gives off far less light than a star like the Sun, thus the contrast between the light of the planet and the light of the star is less. Rather than looking for reflected light (which will also be less if the star is fainter) astronomers study the infrared radiation (heat) from the planet. By choosing very young stars,

the planets will be hotter and hence easier to see.

Unfortunately this method has not identified any planets as yet. A possible planet was found, its brightness in the infrared was as expected and it was the right distance from a young star. When a spectrum of the object was taken, though, it turned out to be a distant K-type star, which through chance had been in the right place and at the right distance to look like a planet. The good thing about this failure was that it showed that this method could distinguish between distant stars and nearer planets. Hopefully next time it will really be a planet.

general astronomy mission than SIM or Kepler. Much of its scientific objective is to determine the distances to huge numbers of stars with very good accuracy and to tell us a lot about the structure of the Galaxy.

As a side project, GAIA will spot every Jupiter-mass or larger planet within 600 light years, which has an orbital period of between 1.5 and nine years. GAIA does not look at stars often enough to identify smaller orbital periods than 1.5 years, but anything larger, up to the nine year lifetime of the mission will be found. Whilst this might not help the search for terrestrial planets, it should give us a fairly complete census of massive planets (especially as current methods are good at finding objects with short orbital periods).

These satellites should provide us, within only a few years, with a good idea of how many Earth-sized planets exist in nearby space. It may only be a few, but these missions could find thousands or tens of thousands if they exist. Once we have located these planets, how can we find out more about them? More importantly, can we find out if they have life on their surface?

3.2   The life zone

Life on Earth has been described as 'a thin green smear'. The diameter of our planet is 12,756 kilometres, but the deepest ocean trenches lie only eleven kilometres below sea level and the average depth of the ocean is just 3.7 kilometres. The highest mountain, reaching nearly nine kilometres above sea level, has a peak at the limits of where the atmosphere can sustain active life forms like ourselves. Looking at Earth from a slightly different perspective, the North Atlantic Ocean is about 4,800 kilometres wide, but has an average depth of just three kilometres. These are the same proportions as a shallow puddle, three centimetres deep and forty-eight metres across. Apart from the fact that life exists at all, there are two particularly striking features of life on Earth which are relevant to the search for life elsewhere in the Universe. First, life started on Earth almost as soon as the planet had cooled four billion years ago; second, life itself has dramatically changed the environment of the thin green smear, helping to maintain conditions suitable for life.

The thin green smear
Life only exists in a tiny band about 15 kilometres thick at the surface of the Earth. This band is shown on the cross section of the Earth here as the outer ring. This gives us some idea of just how small the region that contains the entirety of life on this planet is.

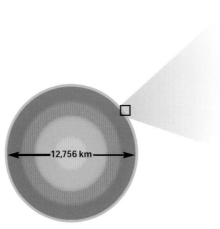

cross section of the Earth

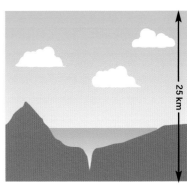

the life zone

Fossil remains of simple forms of life have been found in sedimentary rocks deposited more than 3.8 billion years ago – roughly 500 million years after the Earth and the Solar System formed. This is doubly surprising, because sedimentary rocks (as their name implies) are formed from sediments deposited underwater. In other words, there was liquid water on Earth about four billion years ago. Why should this be surprising? Because according to our understanding of the inner workings of stars, the Sun must have been a little cooler when it was young. This is related to the fact that it has gradually burned hydrogen fuel over the past few billion years. The best astronomical calculations suggest that the Sun would have been so cool four billion years ago that, all things being equal, the Earth would have been a frozen mass, too cold for liquid water to exist. The only explanation is that all things were not equal.

One difference between the early Earth and the way it is today is the composition of the atmosphere. As we have already seen (see page 13), there was no free oxygen in the atmosphere for the first few billion years of the existence of the planet. The original atmosphere, which was produced by 'outgassing' from volcanoes, or deposited on the Earth by comets, or a little of both, when the planet was young, would have been rich in chemically stable gases such as carbon dioxide and methane. These gases are very good at trapping heat near the surface of the Earth – the so-called greenhouse effect (see box page 101). As life began on the planet, photosynthesis reduced the amount of carbon dioxide in the atmosphere and weakened the greenhouse effect, neatly compensating for the increasing warmth of the Sun. The result has been that although the temperature on the surface of the Earth has fluctuated over the aeons, sometimes ice ages, and sometimes warm and wet conditions, the average temperature has been much more stable than it would otherwise have been. This interaction between life and the environment suggests to some that the whole of our planet, both living and non-living components, should be treated as a single entity, which they called Gaia after the Greek goddess of the Earth.

The actual origin of life on Earth remains a mystery. But the speed with which life became established suggests that, at the very least, the raw materials for life were brought to Earth from space. This is not to suggest that any intelligent beings were responsible for seeding the Earth, but in recent years astronomers have detected (by spectroscopy) the presence of many complex molecules, including formaldehyde and amino acids, in clouds like the interstellar cloud from which our Solar System formed (**see box page 105**). The most likely explanation is that large quantities of these so-called organic molecules, the precursors to life, were brought to Earth by comets when the Solar System was young. Again, this has important implications for our search for life elsewhere in the Universe. If interstellar clouds are laced with organic molecules, and these organic molecules are carried down to the surfaces of planets by comets almost as soon as they form, it means that life has a chance to begin on every Earth-like planet that exists in the Universe.

Why only Earth? Why, then, did life not begin on Venus and Mars? The answer seems to be that they were just at the wrong distances from the Sun. Venus is very nearly a twin of the Earth, but is only 70 per cent as far from the Sun as we are. The extra warmth that it would have received when young, combined with the same sort of greenhouse effect that operated on Earth, would have been enough to keep the temperature on Venus above the point where liquid water could exist. As it happens, water vapour is also a very strong greenhouse gas, which would have made the planet even hotter. Venus has been left with a thick atmosphere composed almost entirely of carbon dioxide, creating a greenhouse effect that, combined with the slowly increasing output from the Sun, makes current temperatures at the surface rise above 500 degrees Celsius.

Mars on the other hand is nearly twice as far from the Sun as the Earth is. It is also only half the diameter of the Earth. All the evidence suggests that, long ago, it did have a thick enough atmosphere for the greenhouse effect to keep it warm, and for liquid water to flow. But because the planet was so small its gravity has been too weak to retain this atmosphere, and in spite of the increasing warmth of the Sun the decreasing strength of the greenhouse effect has left it to freeze as the Solar System has matured.

The intriguing feature of this is that if the Earth had a similar orbit to that of Mars, it would retain a sufficiently thick atmosphere to have liquid water oceans. In other words, any Earth-like planet in an orbit between

Everybody has heard of the greenhouse effect, but there is a lot of misunderstanding about how it works. The most important point is that the greenhouse effect is not what keeps a greenhouse warm. In a greenhouse, energy from the Sun penetrates through the glass and warms the contents of the greenhouse. The hot air inside the greenhouse cannot rise by convection and escape because the glass roof acts like a lid – so the greenhouse becomes hotter inside than the air outside.

The atmosphere of the Earth keeps the surface of our planet warm in a different way. Energy from the Sun still gets through to warm the surface of the Earth, but there is nothing to stop convection. The warm surface of our planet radiates energy in the form of infrared heat. Some gases in the atmosphere (in particular carbon dioxide) are very good at absorbing infrared radiation, which they then re-radiate in all directions, with some of it reflected back to the surface and adding to its warmth. This is the atmospheric greenhouse effect. To give you some idea of its importance, the current temperature on the surface of the airless Moon, averaging over day and night, is -18 degrees Celsius. The Earth, at almost exactly the same distance from the Sun, has an average surface temperature of 15 degrees Celsius. The Earth is 33 degrees Celsius warmer than it would be without the greenhouse effect.

The greenhouse effect
Some of the heat radiated from the surface of the Earth is trapped in the atmosphere by 'greenhouse gases' such as carbon dioxide (CO_2). This keeps the Earth's surface much hotter than it would otherwise be.

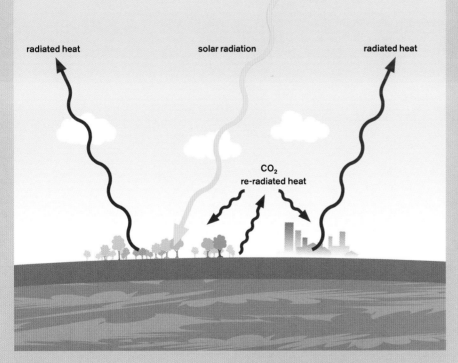

radiated heat

solar radiation

radiated heat

Sun

CO_2
re-radiated heat

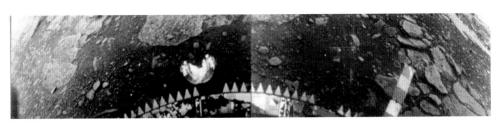

Surface appearances
The surface of Mars (top)
and Venus (bottom)
are both parched and
lifeless. By contrast, the
surface of the Earth
(opposite) is both wet and
lush with vegetation.

the orbits of Venus and Earth to somewhere beyond the orbit of Mars would be an ideal home for life, as we know it. Two of the planets of our own Solar System (Earth and Mars) are in this life zone, even though Mars was just too small to take advantage of the situation. Even Mars, though, might have been habitable today if it had a similar orbit to that of Venus. This is another encouraging indication that homes for life in space are likely to be common.

3.3 Solar System life zones

If we stick to the criterion that water is an essential prerequisite for life as we know it, there are three prime candidates within our own Solar System as homes for life (apart, of course, from the Earth).

Mars Both in science fiction and in science fact most attention has focused on Mars, also known as the Red Planet. This is partly because Mars is our neighbour in space and relatively easy both to study and, in principle, to visit; but also because there is strong evidence that water

Allan Hills meteorite
Found in the Antarctic, this rock that was blasted from Mars may contain the fossilized remains of microscopic life forms. An asteroid impact with Mars threw this rock into space and it eventually landed in the Antarctic.

About one hundred different kinds of molecules have now been discovered in the clouds of gas and dust that lie between the stars. Many of these are very simple indeed – things like water (H_2O) and carbon dioxide (CO_2). But others are much more complex and several of them contain at least ten atoms joined together. One of the most important discoveries is an amino acid called glycine (NH_2CH_2COOH), which is one of the essential building blocks of life on Earth.

Amino acids are the sub-units of proteins, the most important molecules of life apart from DNA itself. Clearly, if molecules as complex as these were brought to Earth by comets soon after the planet formed they could have given a head start to life. It is no coincidence that the atoms that make up a molecule of glycine are nitrogen, oxygen, hydrogen and carbon. These are the most common chemically active elements in the Universe.

Simple compounds containing carbon and hydrogen (CH) and carbon and nitrogen (CN) were discovered more than sixty years ago, using optical spectroscopy. However, the study of interstellar chemistry only came into its own at the end of the 1960s when radio astronomy

techniques were developed which revealed the presence of compounds such as water, ammonia (NH_3) and formaldehyde (H_2CO). Since then, the discoveries have come thick and fast. One of the most dramatic was the identification of ethyl alcohol (C_2H_5OH) which occurs in one single cloud in space (known as Sgr B2) in large enough quantities to make 10^{27} litres of vodka (in everyday language, that is one billion billion billion litre bottles).

In the early 1990s, radio astronomers also began to find evidence of ring-shaped molecules in interstellar clouds. In one of these, pyrene ($C_{12}H_{10}$), one dozen carbon atoms are joined together in a ring with ten hydrogen atoms attached around the outside of the ring. Rings like this can link together to form a very large molecule indeed, known as a polymer. Such polymers are also very important in the chemistry of life.

All of this work shows that biologists, who debated half a century ago how life on Earth could have started from a mixture of simple substances such as carbon dioxide and methane, were barking up the wrong tree. Major steps along the road to life were already taken in the depths of space even before the Earth formed.

Sagittarius B2
The molecular cloud in the Sagittarius B2 complex where stars are forming contains the most complex molecules to be be found in space.

existed in abundance there a long time ago, and may have flowed on the Martian surface relatively recently. Considerable excitement was stirred in 1996 when a team of NASA scientists announced that a meteorite, which had come from Mars, contained evidence of fossilized bacterial life forms. The meteorite had been found in the early 1980s in Antarctica, and there is very little doubt that it really did come from Mars, blasted off the surface of the Red Planet long ago by the impact of a much larger cosmic object. Analysis of the meteorite shows that it must have been ejected off Mars about 16 million years ago, and wandered through space until about 13,000 years ago, when it entered the Earth's atmosphere and fell in Antarctica. Alas, though, there is considerably more doubt about whether the tiny structures revealed by microscopic studies of this piece of rock actually are a sign of life. At the very best, the jury is still out, and we need to send space probes to Mars to search for signs of life.

The evidence for a wet history for Mars comes largely from photographs obtained by space probes orbiting the Red Planet. These show clear signs of features such as gorges and dried-up riverbeds that can only have been carved by flowing water. When these features were first studied, thirty years ago, it seemed that they were very old and that water had ceased to flow on Mars billions of years ago, having been frozen into ice below the surface. Even if life had started early in the history of the planet, there was little likelihood that it would survive today. More recently, however, there have been claims that there is evidence of very recent water flows on the surface of the planet. Higher resolution images from the space probe Mars Global Surveyor revealed channels on the sides of some Martian craters which suggest that water has flowed there very recently. To an astronomer, 'very recently' still means that the channels could be one or two million years old, but equally they might only be a few years old. These channels, seen in the northern hemisphere are, strangely, on the southern (that is pole-facing) inner slopes of the craters. These are amongst the coldest places on Mars. However, because Mars lacks a large stabilizing Moon, it wobbles on its axis much more than the Earth does. About four million years ago it was temporarily tipped over so far that the north-facing crater slopes would have received the full heat of the Sun throughout the summer (just as the Sun never sets in summer at

Mars
Details of the Martian surface can clearly be seen in this stunning image of the Red Planet taken by the Mars Global Surveyor. Mars is one of the better options in the Solar System for finding life, or at least evidence of extinct life.

the north pole of the Earth). At the same time, it is likely that evaporation from the polar caps would have thickened the atmosphere of the Red Planet, enhancing the greenhouse effect and also encouraging any sub-surface ice to melt. The fact that the atmosphere of Mars is in chemical equilibrium and consists almost entirely of carbon dioxide is not an encouraging sign for anyone hoping to find life there. But because there is water on Mars and because the planet is so accessible, it is a prime target for probes designed to search for life (see box page 109).

Europa The second location beyond Earth where space scientists are already planning to search for signs of life is Europa, one of the moons of Jupiter, five times as far from the Sun as the Earth is. Europa is slightly smaller than our own Moon, with a diameter of 3,139 kilometres. The space probe Galileo sent back stunning pictures of Europa, which showed it from a distance as having the appearance of a smooth glass-marble covered in vein-like cracks. Close-up images show that the surface consists of what seems to be a layer of cracked ice on top of liquid water, which has flowed through the cracks and frozen. Superficially, this is very like the appearance of the ice-covered Arctic Ocean, but the ice crust on Europa is likely to be significantly thicker. The reason why Europa may be warm enough for liquid water to exist, even though it is so far from the Sun, is that this moon is constantly being stretched and squeezed by the tidal influence of its giant parent planet, Jupiter. This squeezing generates heat in the rocky core of the moon, which could maintain a salty, liquid ocean with a temperature probably just below the freezing point of freshwater. Europa's ocean could be as much as one hundred kilometres deep and covers the entire moon. The ocean in turn would be covered by a layer of slushy soft ice up to twenty kilometres thick in which convection and partial melting brings liquid water to the surface. The cracked layer of surface ice is itself a few kilometres thick and has a temperature of about -170 degrees Celsius, but acts as an insulating blanket around the moon trapping what little heat there is in the underlying ocean. The next step in the search for life on Europa will come when NASA's Europa Orbiter probe is launched – at present the launch is scheduled for 2006, but this date has not yet been finalized.

The Beagle Lander
Due to touch down on the
surface of Mars in the
latter half of 2003, the
Beagle Lander will search
for evidence of life on
Mars.

The next mission to search for life on Mars will be launched in June 2003 by a Russian Soyuz rocket, and will arrive at Mars in December of the same year. The main mission, called Mars Express, is a European Space Agency (ESA) project designed to put a probe in orbit around Mars. But the probe will also carry a small landing vehicle, called Beagle 2 in honour of HMS Beagle, the ship in which Charles Darwin travelled around the world. The mass of the lander is just under thirty kilograms, about the same as the baggage limit on many airlines, and represents what the British-led team responsible for the design calls 'the most ambitious science payload to systems mass ratio ever attempted'. The design of the spacecraft is like a clamshell, which will reach the surface after being ejected from the Mars Express orbiter and slowed down using parachutes. The impact it will have to withstand is about the same as if you pushed a laptop PC off a chair onto the floor. The spring-loaded clamshell will then be able to pop the Beagle Lander open regardless of the orientation it ends up in. Its instruments will then begin testing for the presence of water, carbonate minerals of the kind that might have been laid down underwater, and the kind of organic materials that would have been produced by life. The primary mission is intended to last for 188 Martian days (each

almost the same length as a terrestrial day), and the team hopes that the probe will remain operational for a full Martian year (twice as long as an Earth year).

Beagle 2 will search for life on Mars in two ways. First it will search for methane in the atmosphere. Like oxygen, methane does not last long on its own (at reasonably high temperatures anyway) and so any methane in the atmosphere will have been produced recently. In general, looking for methane is not a good way of finding life, as vulcanism can produce methane. However, on Mars we do not think that there is any active volcanoes and so methane could well be a sign of life. Beagle 2 will also burn rocks, slowly heating them to search for signs of life. Organic materials burn at lower temperatures than rock (generally 100–300 degrees Celsius, whereas rock melts at more than 500 degrees Celsius). As well as looking for low temperature organics Beagle 2 will also analyse the isotopes of carbon. Most carbon has an atomic weight of 12 (atoms are weighed by how much heavier than hydrogen they are), but life takes up other isotopes, carbon-13 and csvin-14, from the environment in a way which alters the ratio of these isotopes in living things compared with the non-living world. Beagle 2 will look for a higher-than-average composition of carbon-12, which would be a sign that life had formed there.

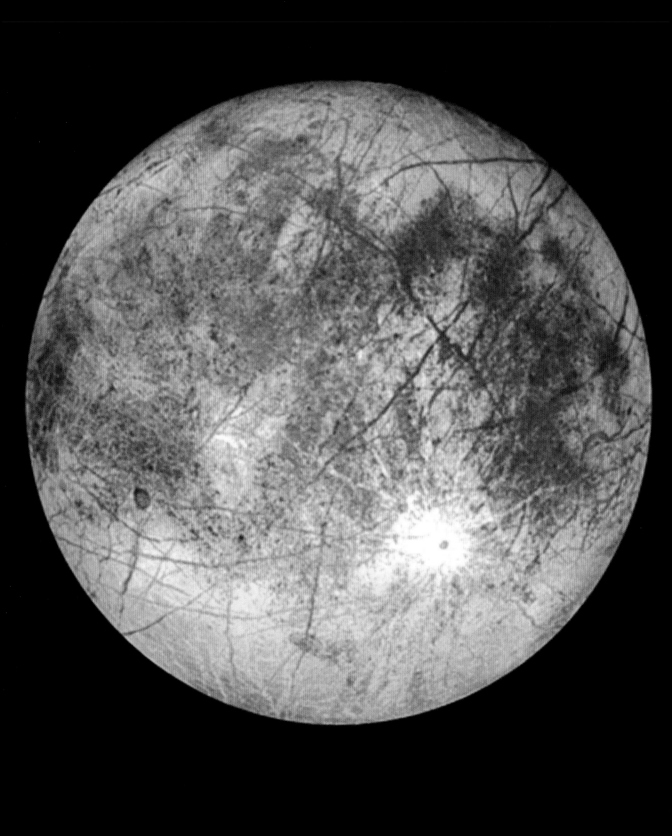

Europa
An ocean of liquid water is thought to exist below the frozen surface of Europa. The water is kept liquid by the heat created by Jupiter's gravity, which massages the core of this moon.

After that, there are plans to drop a probe onto the surface of Europa to penetrate the icy crust and find out just what does lie beneath.

An ideal site for testing the Europa probe is in Antarctica. Buried under 3.7 kilometres of ice is the recently discovered Lake Vostok. It is about 500 metres deep and covers an amazing 14,000 square kilometres. Lake Vostok has been isolated, under the ice, from the rest of the planet for possibly one million years. Hydrobots (underwater robots) are currently being built to search the lake. These hydrobots are the precursors of the type of probe that it is hoped will be sent to Europa to search for life under the icy surface.

Titan The other possible home for life in the Solar System is something of a long shot. The planet Saturn is nearly ten times further away from the Sun than the Earth is. However, it has a moon, Titan, with a diameter of 5,150 kilometres and a thick atmosphere, mainly composed of nitrogen, but also containing methane. The pressure of this atmosphere at the surface of Titan is 1.6 times the pressure of the atmosphere at sea level on

Deep-sea vent
Volcanic fissures found deep below the ocean surface on Earth provide an outlet for heat and chemicals, which allow life to survive despite the lack of sunlight. Similar vents may exist on the floor of Europa's 'ocean'.

Cassini
This artist's impression
of Cassini shows this
NASA probe (scheduled
to reach Titan in 2004)
dropping the Hygens
lander into the
atmosphere of Saturn's
largest moon, Titan.
Hygens will sample the
atmosphere and surface
of Titan to discover their
chemical compositions.

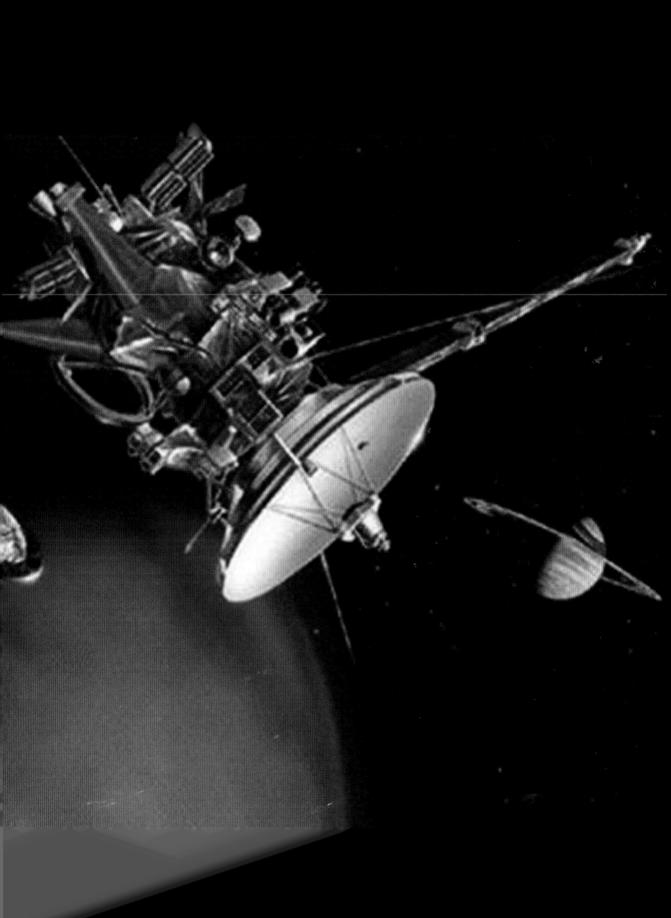

the Earth, but the temperature on the surface of Titan is –180 degrees Celsius. Titan has been described as a smaller version of the early Earth, put currently in a deep freeze. In about five billion years time, when the Sun reaches the end of its life on the main sequence and becomes a red giant, the Earth will be fried and become unsuitable for life. However, Titan will be warmed and could become even more Earth-like.

Some optimists hope that there may already be forms of life on Titan, inhabiting lakes or oceans of liquid methane and enjoying the methane rain falling from the clouds. More realistically, it is hoped that by studying the atmosphere of Titan today astro-biologists can learn more about the conditions under which life emerged on Earth, and thereby improve their understanding of the probability of finding life elsewhere in the Milky Way.

To that end, the Cassini spacecraft, launched at the end of 1997, will drop a probe into the atmosphere of Titan to find out more about the chemical processes at work. Cassini has taken a rather roundabout route to Titan and will arrive in 2004 to deposit the Hygens Lander probe. It

Titan from Earth
Captured by the Canada-France-Hawaii telescope, this infrared image of Titan penetrates this moon's dense atmosphere, revealing different features, which might be oceans and continents.

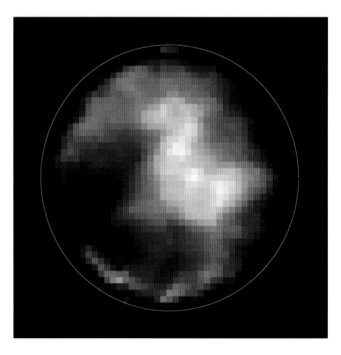

will then investigate the rest of Saturn's moons, as well as the rings of Saturn and, of course, the planet itself.

Although you need to send a probe to Titan to study its chemistry, Earth-based telescopes are still providing surprises about this moon. Using the Canada-France-Hawaii telescope based in Hawaii, in the summer of 2000 astronomers reported that they had obtained images of the surface of Titan at infrared wavelengths. Infrared radiation penetrates the clouds of Titan and reveals a shiny region about the size of Australia, which is thought to be a range of icy mountains. At -180 degrees Celsius, ice is as strong as granite, and could easily support the weight of such a huge formation. But because the mountains are made of ice, they are probably not permanent features. Wet air blowing up from the methane oceans will produce a methane rain that erodes the features and explains why the shiny surface is exposed. However, whatever their origin, the discovery of mountains of ice on Titan is further proof that the substances required for life (even within our own Solar System) are not unique to the Earth.

3.4 Life zones around other stars

We have already seen where in the Solar System the right conditions for life might exist. The main area is the habitable zone from slightly closer than the Earth's orbit to Mars' orbit. In this region, large enough planets with atmospheres may have the right temperatures for liquid water to exist. The most sensible places to look for life-bearing planets would appear to be these habitable zones. Habitable zones have two parameters: their distance from the parent star and their duration. Both of these parameters are determined by the spectral type (or mass) of the parent star. The temperature of the star determines the distance of the habitable zone from the central star. The hotter the star, the further away the habitable zone is. In the diagram on page 116 we see how the distance from a star at which life is sustainable depends on the star's spectral type (i.e. its mass) which determines its temperature. Small, low temperature M-type stars have a habitable zone that is very close to the parent star, close to the orbit of Mercury. A G-type star, like the Sun, has a habitable zone at about the Earth-Sun distance. As the central star gets hotter the habitable zone moves further away.

The life zone

In this graph, the yellow band shows how the life zone moves further away from stars as the stars become bigger and hotter. Anything to the left of the red line is tidally locked. As can be seen, the remaining zone in which habitable planets might possibly exist is fairly small.

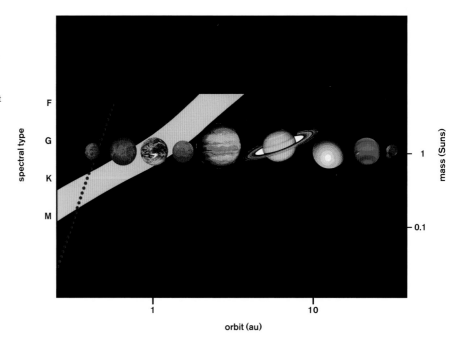

The limits of habitable zones are determined by the presence of liquid water on a planet's surface (**see page 125**). This means that the temperature of the surface must be between about 0 and 100 degrees Celsius. This does not mean that a planet is absolutely required to be in the habitable zones we have indicated above. However, this is the case for a planet very much like the Earth. If we wanted to form a planet with the right temperature there are two main features of the planet we can play with. If the planet is too close to the star we can lower the temperature by making the planet reflect heat away (using cloud cover, ice caps – anything light coloured and reflective) whilst keeping the greenhouse effect low. If the planet is too far away we can increase the temperature by having a stronger greenhouse effect (normally, a thicker atmosphere) and having little to reflect sunlight away. It has been found in the case of Saturn and Europa that tidal massaging (where the gravity of a planet pushes and pulls a moon, causing it to change shape slightly) has a warming effect on the smaller partner.

Where beyond the solar system should we look? Which types of stars have good habitable zones around them? Whilst all stars have a zone around them where the temperature criteria can be fulfilled, many stars have other factors which make them hostile to life. Stars more massive than two or three solar masses just do not live long enough for advanced life to develop. The lifetime of their habitable zones, a few hundred million years, maybe a billion years at most, is just too short for life. This still leaves us with most of the stars in the Galaxy. However, unfortunately we must probably rule out 75 per cent of all stars, as they are too small. M-type dwarfs smaller than about 0.3 solar masses have a habitable zone that is very close to the star because they are very cool (in stellar terms). These orbits would be so close in fact, that a planet in the habitable zone could have a year as short as two weeks. Any planet inside this habitable zone would be tidally locked to the star. This means that they would always have the same side facing the star, just like the Moon relative to the Earth or Mercury relative to the Sun.

Tidal forces occur because the gravitational force of a star is slightly less on the far side of a planet than on the near side (because it is further away). This acts to slow down the rotation of the planet very slightly. Over the course of millions or billions of years it will make the planet rotate about its axis at the same rate as it orbits the star, so it will always keep one side facing the star. If we wait long enough the Sun will eventually have the same effect on the Earth.

This tidal locking would mean that one side of the planet would be constantly heated and would boil, while the other side of the planet would never see the star and would freeze. Some recent simulations have shown that this might not always be the case; planets that are tidally locked may have very complex atmospheres that circulate heat from the hot side onto the cold side to try and keep the temperature of the whole atmosphere relatively stable. Life might well be able to start on such planets although, from our point of view, it may not be very hospitable.

The other problem with very small stars is that the habitable zone is so close that it may be in danger from stellar flares. These are energetic bursts of ionized particles that are emitted by stars. The Sun produces flares that can disrupt communications across the globe and even cause power blackouts. Many small stars have flares that are more frequent and more violent than those of the Sun. These flares might sterilize the planets close to the star in the habitable zone every so often, killing any life that may have got started.

This means, once again, that the best types of star for allowing life to develop are stars like our Sun. The habitable zone lasts long enough for life to develop and is far enough away not to be effected by solar flares or tidal locking.

However, habitable zones do move. As we have mentioned previously (section 1.2), the Sun was fainter and cooler in the past than it is now, and this means the habitable zone was closer to it. The Sun is evolving, very slowly and recent calculations have shown that in two or three billion years time the habitable zone around the Sun will move outwards towards the orbit of Mars, and the Earth will actually move out of the life zone. This could cause a runaway greenhouse effect on the Earth and kill all life on the planet, turning it into another Venus.

3.5 Signs of life

Life forms like ourselves need a lot of energy to run about and do interesting things, and also need a lot of energy to run a complex organ like the brain. Plants need a much smaller supply of energy, because they do not need energy to run about or to think. The energy that enables us to do interesting things comes from chemical reactions that involve oxygen. A little thought shows how dangerously reactive oxygen is – without oxygen there would be no forest fires or burning oil wells, and no vehicles propelled by internal combustion engines that draw on material from the atmosphere to make up their internal fires. The chemistry that allows animals to do interesting things is just a slowed-down form of combustion, and the reason why we do not usually think of oxygen as being dangerous and potentially harmful is simply that familiarity of it has bred contempt.

Oxygen is so reactive that if it were not being continuously produced today on Earth by plant life, all of the oxygen in the atmosphere would disappear within a couple of hundred years, used up in forest fires, rusting, and other such processes which lock oxygen away in stable compounds. By studying the nature of rock strata laid down over the past four billion years, geologists are able to tell that there was no free oxygen

Forest fire
Conflagrations such as forest fires are only possible on a world that contains lots of free oxygen in its atmosphere Such conflagrations are an example of the energy that burning oxygen can provide naturally.

in the atmosphere until about two billion years ago. Earlier life forms used the energy from sunlight to build up the complex compounds they required from raw materials such as carbon dioxide and hydrogen sulphide. But then, some bacteria evolved that used a different kind of photosynthesis, in which they extracted the hydrogen they needed from water instead of from hydrogen sulphide. Water is simply a compound of hydrogen and oxygen, but these photosynthesising bacteria had already extracted all the oxygen they needed from carbon dioxide. So to them, oxygen was a waste product, which could be thrown away.

For about one billion years, the excess oxygen produced in this way was locked up in compounds of iron before being released, because to the organisms at that time oxygen was a dangerous poison. Later still, around two billion years ago, bacteria evolved that could cope with the presence of free oxygen in the environment. This gave them a double advantage over their rivals. First, they could throw their spare oxygen straight into the atmosphere, without going through the stage of combining it with iron; second, the oxygen released in this way poisoned many of their

Oxygen powered

The use of oxygen by dolphins and other animals on Earth allows them rapid movement and an energetic lifestyle.

Terrestrial spectra

The different spectra of the terrestrial planets are markedly different. Earth shows the distinct signatures of both liquid water and oxygen, while these life sustaining ingredients are lacking on both Venus and Mars.

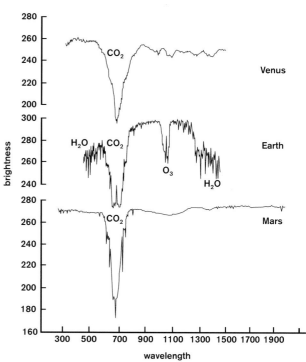

1	2	3	4	5	6	7	8	9	10	11	12	13	14	15	16	17	18
1 **H** 1.0079																	2 **He** 4.0026
3 **Li** 6.941	4 **Be** 9.0122											5 **B** 10.811	6 **C** 12.011	7 **N** 14.007	8 **O** 15.999	9 **F** 18.998	10 **Ne** 20.180
11 **Na** 22.990	12 **Mg** 24.305											13 **Al** 26.982	14 **Si** 28.086	15 **P** 30.974	16 **S** 32.066	17 **Cl** 35.453	18 **Ar** 39.948
19 **K** 39.098	20 **Ca** 40.078	21 **Sc** 44.956	22 **Ti** 47.867	23 **V** 50.942	24 **Cr** 51.996	25 **Mn** 54.938	26 **Fe** 55.845	27 **Co** 58.933	28 **Ni** 58.693	29 **Cu** 63.546	30 **Zn** 65.39	31 **Ga** 69.723	32 **Ge** 72.61	33 **As** 74.922	34 **Se** 78.96	35 **Br** 79.904	36 **Kr** 83.80
37 **Rb** 85.468	38 **Sr** 87.62	39 **Y** 88.906	40 **Zr** 91.224	41 **Nb** 92.906	42 **Mo** 95.94	43 **Tc** (98)	44 **Ru** 101.07	45 **Rh** 102.91	46 **Pd** 106.42	47 **Ag** 107.87	48 **Cd** 112.41	49 **In** 114.82	50 **Sn** 118.71	51 **Sb** 121.76	52 **Te** 127.60	53 **I** 126.90	54 **Xe** 131.29
55 **Cs** 132.91	56 **Ba** 137.33	57-71 *	72 **Hf** 178.49	73 **Ta** 180.95	74 **W** 183.84	75 **Re** 186.21	76 **Os** 190.23	77 **Ir** 192.22	78 **Pt** 195.08	79 **Au** 196.97	80 **Hg** 200.59	81 **Tl** 204.38	82 **Pb** 207.2	83 **Bi** 208.98	84 **Po** (209)	85 **At** (210)	86 **Rn** (222)
87 **Fr** (223)	88 **Ra** (226)	89-103 #	104 **Rf** (261)	105 **Db** (262)	106 **Sg** (263)	107 **Bh** (264)	108 **Hs** (265)	109 **Mt** (268)	110 **Uun** (269)	111 **Uuu** (272)	112 **Uub** (269)		114 **Uuq** ‡		116 **Uuh** ‡		118 **Uuo** ‡

*Lanthanide series	57 **La** 138.91	58 **Ce** 140.12	59 **Pr** 140.91	60 **Nd** 144.24	61 **Pm** (145)	62 **Sm** 150.36	63 **Eu** 151.96	64 **Gd** 157.25	65 **Tb** 158.93	66 **Dy** 162.50	67 **Ho** 164.93	68 **Er** 167.26	69 **Tm** 168.93	70 **Yb** 173.04	71 **Lu** 174.97
#Actinide series	89 **Ac** (227)	90 **Th** 232.04	91 **Pa** 231.04	92 **U** 238.03	93 **Np** (237)	94 **Pu** (244)	95 **Am** (243)	96 **Cm** (247)	97 **Bk** (247)	98 **Cf** (251)	99 **Es** (252)	100 **Fm** (257)	101 **Md** (258)	102 **No** (259)	103 **Lr** (262)

The Periodic Table
This table sets out all the chemical elements in order of their atomic weight starting with the lightest element, hydrogen. Every element more massive than Beryllium (Be), the fourth element in the table, has been built up by nuclear processes within stars. Many fairly common elements such as carbon, oxygen and nitrogen are vital for life.

rivals. However, life is very adaptable and it is no coincidence that animal life evolved soon after a reservoir of oxygen had been established in the atmosphere to provide the fuel for their more active lifestyles.

If you were to study the atmosphere of the Earth from space, as a visiting alien might from a spacecraft approaching the Solar System, it would be obvious that our planet is a place where interesting things are happening. In chemical terminology, the presence of so much oxygen means that the atmosphere is far from equilibrium – like a ball balanced on top of a hill, all things being equal it would 'roll down' to an equilibrium state. By comparison, to our hypothetical alien visitor it would be very obvious that nothing interesting (in terms of active life forms like ourselves) exists on Venus or Mars. Both of those planets have atmospheres mainly consisting of carbon dioxide, a very stable compound, which is in chemical equilibrium with all the chemical compounds (such as carbon dioxide), at the bottom of the energy hill. From this evidence alone it would be clear that active life forms with energy intensive brains do not exist on Venus or Mars.

The reactions that make nitrogen can only take place inside stars where there is already some carbon and oxygen, which has been built up by combining alpha particles.

The process begins when a proton collides with a nucleus of carbon−12 (containing six protons and six neutrons). This creates an unstable nucleus (nitrogen−13), which emits a positron and becomes a nucleus of carbon−13.

When a second proton collides with this nucleus, it is converted into nitrogen−14; but this is not the end of the story.

The addition of a third proton can convert the nitrogen−14 nucleus into oxygen−15, which is also unstable and ejects a positron as it converts into nitrogen−15. If yet another proton combines with this nucleus, it ejects one alpha particle, leaving behind a nucleus of carbon−12, identical to the one that started the cycle.

The net effect is that four protons have been converted into one alpha particle, with energy being released along the way. This so-called carbon cycle operates in the heart of stars where the temperature is above 21° Celsius. It is important in the manufacture of nitrogen because whatever the proportions of oxygen and carbon that you start with (at the very beginning of the process, there will be no nitrogen at all) you always end up with the same ratio of the three main elements involved.

The different steps of the cycle proceed at different rates, and, in particular, the reaction that converts nitrogen−14 into oxygen−15 proceeds very slowly. Therefore equilibrium is achieved when there is 5.5 % carbon−12, 0.9 % carbon−13, 93.6 % nitrogen−14, and just 0.004 % nitrogen−15.

As well as providing the energy to keep massive stars shining, the carbon cycle is a very efficient way of making nitrogen.

There is virtually no other way for stars to make nitrogen, and it is absolutely certain that every atom of nitrogen in our bodies was made by the carbon cycle operating inside a hot, massive star which later died and scattered it's ashes across the Universe.

The carbon cycle
In the heart of many stars the collision of subatomic particles builds up large amounts of nitrogen. This element, which is a vital ingredient of life, cannot be formed in any other way.

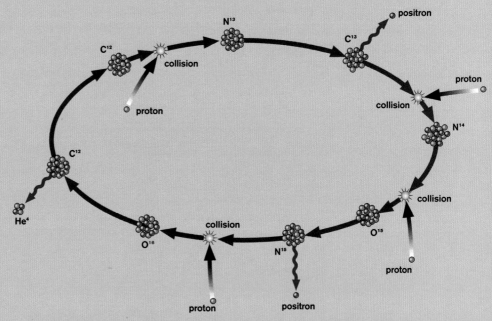

An analogy is the contrast between a car powered exclusively by electricity from solar panels, and one powered by petrol. The electric car does work, but it cannot move as quickly or carry such a large load as the internal combustion powered vehicle. It is easy to see why – fossil fuels such as petrol are essentially stored-up sunlight, energy from the Sun converted into a chemical form by photosynthesis, and it takes an awful lot of sunlight to make a gallon of petrol.

If it were possible to obtain spectra of the light from distant planets, then it would be just as easy to pick out the planets most likely to harbour active life forms like ourselves, just as we can do when we compare Venus, Mars and the Earth – but, what a big if!

Apart from the carbon, hydrogen and oxygen that we have already mentioned, life as we know it requires one other element in particular: nitrogen. These four elements together are so important for life that they are often referred to by an acronym (CHON). No less than 65 per cent of the mass of a human body is made up of water (in other words, hydrogen and oxygen), but half of the remainder is carbon, one quarter is oxygen,

Solar power versus combustion
In current automotive technology, combustion provides far more energy than collecting sunlight. It is difficult to imagine a time when a solar powered car could beat a formula 1 car.

and just less than 10 per cent is nitrogen. All of the hydrogen dates back to the Big Bang during which the Universe was born about thirteen billion years ago. The other elements have been manufactured inside stars by nuclear fusion reactions and scattered through space when the old stars die, sometimes in a very spectacular supernova explosion.

Why should these particular elements be so important for life forms like ourselves? It may simply be because these are the elements that were available for life to work with. The kind of interstellar clouds from which stars, planets and ourselves are made are almost entirely composed of hydrogen (70–75 per cent) and helium, with everything else making up only about one per cent. Taking that one per cent alone, oxygen is the most common element, carbon is the second most common, and nitrogen is the third. Roughly speaking, for every ten nitrogen atoms there are forty atoms of carbon and seventy atoms of oxygen. Even though these are just trace quantities compared with the thousands of atoms of helium and tens of thousands of atoms of hydrogen, they are the basis of all the interesting chemistry in the Galaxy.

Water In addition to free oxygen in the atmosphere, which makes it possible for active animals with large brains to exist on Earth, our planet is more than half-covered by oceans of liquid water. Water is an essential requirement for life as we know it, and water is, of course, made up of hydrogen and oxygen – the two most common chemically active elements in the Universe. (Although there is a lot more helium than oxygen, helium does not take part in chemical reactions, which is one reason why it was not identified on Earth before 1895.) If you find a planet with water, it is a likely home for life, although not necessarily animal life; if you find a planet with free oxygen as well as water, then there is a good chance that animal life can exist on that planet.

There is no mystery about the process by which the elements are made. As we have explained, the Sun counteracts the inward pull of gravity by converting hydrogen into helium in its interior, and releasing energy as a result. Hydrogen is the simplest element there is, and this is all the Sun has to do to prevent itself from collapsing. When a star like the Sun has used up all of its hydrogen, it has to generate energy in other ways.

Why is water considered so vital for life? All terrestrial life seems to depend upon liquid water. Even in Antarctica bacteria are found where a small supply of liquid water is available amongst the vast frozen deserts.

The major reason why water is required is that cells require a solvent – a medium into which chemicals can be dissolved and in which chemical reactions can occur. Water is by far the best liquid available for this. Indeed it is the closest thing to the 'universal solvent' that we have yet found, searched for by the ancient alchemists.

As well as the amazing ability of water to dissolve other substances, it has two other properties that make it different from almost any other molecule. First it is polar, that is one side of the molecule has a slightly different magnetic charge to the other. This helps define the shape of amino acids (long chains of molecules) on which life is based. A very special property is that solid water (ice) is less dense than liquid water, which means that ice floats. Thus water freezes from the top downwards leaving a liquid under the ice which helps to thaw the ice when the water melts. If it were not for this property all the ice in the oceans would sink to the bottom where sunlight could not reach to melt it.

The closest liquid to water is ammonia. It shares the ability of water to behave as an almost universal solvent, but unfortunately does not share many of water's other properties. Of course, we are basing our ideas of the importance of water on the one example that Earth-based life needs water. In this we may be totally wrong. It does seem though, that 'life as we know it' would need liquid water.

The more massive the star is, the easier this is, and all the variety of elements except for hydrogen and some original helium have been made inside stars as a result. However, concentrating only on the ones that are most interesting for life, as we know it, it is very easy to see where the elements come from, through further stages of nuclear fusion. The nucleus of a carbon atom basically consists of three alpha particles (helium nuclei) that have fused together. Adding a fourth alpha particle makes oxygen. Nitrogen is made by a slightly more subtle process involving both carbon and oxygen which provides the energy source inside hot, massive stars where the temperature in the centre rises to 20 million degrees Celsius (see box page 122).

The important message to take away from this understanding of how stars work and how the elements are made is that we are made from the most common kind of material available. This makes it very likely that other forms of life in our Galaxy are made from the same kind of material that we are, and suggests that the obvious place to look for life forms like ourselves is indeed on planets like the Earth.

3.5 Looking for life

Space-based astrometry missions such as GAIA and SIM (see section 3.1) may provide us with information on Earth-sized planets around nearby stars if they exist. The size and period of the wobbles of these new planets will tell us the mass and orbits, but that is all they can tell us.

If we find Earth-sized planets around other stars the crucial questions we would like to find an answer for are: What are conditions like on the planet's surface? Is there liquid water? And, most importantly, is there evidence of life? To find the answers to these questions we must look for a planet finding technique that is more advanced than astrometry or transit detectors.

As we have already mentioned (see section 3.1), observing planets directly is very difficult. Stars are very bright and planets are very faint in comparison. The other problem is that planets, and terrestrial planets in particular, are expected to be close to their parent star. In addition to looking for a very faint planet, we are looking for it at a distance of only milliarcseconds from the star. Such requirements imply that the direct imaging of planets is beyond the current capabilities of telescopes.

In the near future space-based telescopes, capable of imaging Earth-like planets directly, could be built. Not only could they be built, but also one of these telescopes could be operational within ten to twenty years.

Both NASA and ESA are at fairly advanced stages of planning for these telescopes. NASA has named their mission Terrestrial Planet Finder (TPF), while ESA is deciding between Darwin and the InfraRed Space Interferometer (IRSA). Both of these missions are very similar and some talks have been held between NASA and ESA aimed at a collaborative mission favoured by administrators in order to reduce the costs of developing and launching two separate missions.

Darwin and TPF will operate in the infrared region of the electromagnetic spectrum. The reason for selecting the infrared region is because planets, especially terrestrial planets are brightest at these wavelengths. The wavelengths of the light that will be observed are of the order of microns (millionths of a metre), which is tens to hundreds of times longer in wavelength than visible light. The wavelength of light from an object

Comet Hyatake
Comets are thought to have been an important source of water and other useful chemicals to the young Earth.

The Very Large Array
This assembly of Earth-based radio telescopes in Socorro, New Mexico uses the interferometric techniques Darwin and TPF will use in space.

Darwin
Amongst the most advanced projects to launch multiple telescopes into space to look for extrasolar planets capable of supporting life is Darwin. This artist's impressions of Darwin shows an array of six space telescopes, as well as a central hub to collect the data from each telescope, and a communications satellite to beam the data back to Earth.

depends on its temperature. The Sun with a surface temperature of 6,000 degrees Celsius emits most light in the middle of the visible spectrum at a wavelength of one fifty-millionth of a metre. The Earth is much cooler with a temperature of only about 15 degrees Celsius and emits most of its light at a wavelength of a few microns. At this wavelength Earth is the brightest planet in the Solar System, even brighter than Jupiter (even though Jupiter is much bigger, it is cooler as well). In this region of the infrared terrestrial planets stand out like the proverbial sore thumb.

Interferometry The sophisticated technology behind these missions is called interferometry (the same idea behind SIM). The main advantage of an interferometer is that it allows several small telescopes to be grouped together to make one larger telescope (see box page 135).

This technique has been used in radio astronomy for many years. A set of radio dishes, kilometres (or even thousands of kilometres) apart, can be set up to act as one large radio telescope. The collecting area is the size of all the dishes added together. The telescope does not collect any more light than

Light can be thought of as a wave – a series of peaks and troughs, much like waves on the water. Two waves from different sources can mix. If two peaks or two troughs are at the same point they add together and make a bigger peak or deeper trough. If a peak and a trough are at the same point then they cancel each other out.

This process forms the interference pattern in the picture – bright lines appear where two peaks or two troughs combine and dark lines appear where they cancel out. Thus, if the distance between the two telescopes is known to a very high accuracy then they can be combined to act like one giant telescope with a diameter of hundreds of meters.

Even better, the light from the star can be made to cancel-out between telescopes adding peaks and troughs. Light from the planets doesn't cancel as they are in a slightly different position on the sky. This means that the star which usually outshines the planets is removed and only the planets remain.

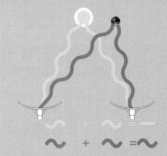

each individual telescope on its own, but the resolution can be incredibly good. The resolution is determined by the separation of the telescopes and so radio telescopes are able to obtain resolutions of milliarcseconds.

The limitation of this technology is that the signals from each telescope must be timed very accurately or, normally, the distance from each telescope to the main receiving station must be known very accurately. This distance must be known more accurately than the wavelength of the light being received. In the case of radio astronomy, where the wavelength is of the order of metres or centimetres, this is not too difficult but for infrared or visible light, where the wavelength is of the order of millionths of a metre, it is far more difficult.

The great strength of this technique is that the telescopes can be set up so that light from one object will not be seen whilst from other objects it will be strongly visible. When searching for planets the central star can be removed from the image – leaving just the fainter planets behind!

Darwin and TPF will have the ability to spot planets like Venus, the Earth and Mars in orbit around a central star (Mercury is too small and close to the central star to be seen).

However, observing at infrared wavelengths, even from space, does pose problems. Most of the radiation is prevented from reaching the Earth's surface by the atmosphere (in most cases this is for the best as far as we are concerned). Normally putting a satellite in orbit would be enough to see all of the radiation, but in the infrared wavelength band that Darwin and TPF will be using the inner Solar System is not a good place to be either.

The whole of the inner Solar System contains a feature called zodiacal dust. This is small pieces of comet debris boiled off from comets as they approach the Sun. The dust is mainly located in the plane of the Solar System, in the band of constellations that the Sun passes through as we look from the Earth – the zodiac. The zodiacal dust causes a problem because the sizes of the dust grains are about the same as the wavelength of the infrared radiation band we want to observe. When the wavelength of radiation and the size of a dust grain or gas molecule it is passing through are about the same the dust or gas scatters the radiation very effectively.

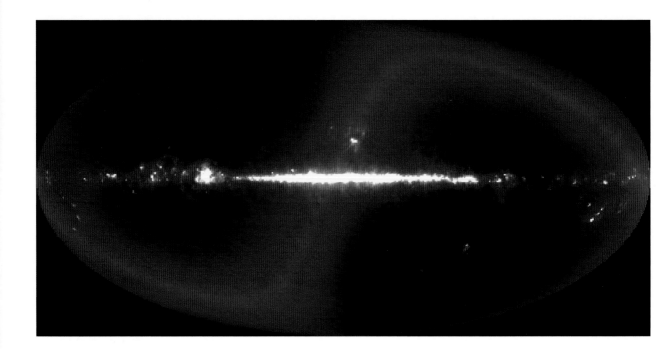

Zodiacal Dust
Photographed by an
infrared camera, this
image shows the Milky
Way as a bright band
running across the centre
of the picture. The
dramatic S-shaped band
that appears to engulf the
Milky Way is in fact merely
the light produced by
grains of dust in the Solar
System's asteroid belt.

To get around this problem the plan is to send both missions out of the inner Solar System to near the orbit of Jupiter where the amount of zodiacal dust is far lower. The satellite will sit at the Lagrange point, where the gravitational pull of Jupiter and the Sun are equal. Such balance points are in the same orbit as the parent planet (in this case, Jupiter) at a certain distance in front of and behind the planet in its orbit. Any object placed at these points will stay that distance from the planet, orbiting the Sun at the same speed as the planet, which means that the telescope would sit there until the mission was over.

The designs of the satellite for Darwin and TPF are also very similar. Both will probably consist of a group of six separate satellites that must fly in formation. All but one of these satellites will be infrared telescopes, each about 1.5 metres in diameter. There will also be a central satellite that will combine the information from each telescope and relay it to Earth.

One of the main problems of observing in the infrared is the contribution of excess heat. The satellites themselves emit infrared radiation of the same wavelength as the telescopes are observing. For this reason, the telescopes must be cooled and kept shielded from the heat of the Sun. Using liquid gases (probably nitrogen) which keep the heat sensitive instruments at a temperature of many degrees Celsius below zero does this cooling. The coolant supply is limited, and it is this limitation that determines the lifetime of the mission.

One of the biggest technological challenges of the mission is getting several spacecraft to fly in formation. The positions of each of the telescopes relative to the central satellite need to be known with high accuracy. The distance between spacecraft may be more than 100 metres but must be known with an accuracy of less than one millimetre. All of this will have to be done more than 600 million kilometres from the Earth.

Darwin and TPF, or (most probably) a combined mission using the best features of each proposal, could be launched in the early part of the next decade and will have a lifetime of about five or six years. Both Darwin and TPF have very similar mission plans. They will be able to observe terrestrial planets around a few hundred stars within a distance of fifty or-so light years. Depending on the success of earlier missions like Kepler and SIM (see section 3.1), target stars with known or suspected terrestrial planets can be selected. If no candidates are available, target stars like the Sun will be chosen, preferably heavy-element rich, fairly old, stable, and single stars will be selected. Darwin currently has a target list of approximately 200 nearby stars.

For each of these stars observations of a few tens of hours would be enough to get a picture like we saw above. Images taken over the course of weeks and years would show the movements of the planets, which would provide us with all the basic information about the orbits and inclination of the system, giving the masses of the planets. Zodiacal dust can hamper the search for other planetary systems. Just as it scatters the light coming into our inner Solar System, it can also scatter the light leaving other solar systems, thus dimming the planets and making them more difficult to see. The level of zodiacal dust in our Solar System does not in itself cause us insurmountable difficulties in detecting extrasolar planets, but systems with considerably more zodiacal dust (for example, young planetary systems, or systems which have recently seen the break-up of a giant comet) are likely to obscure some planets.

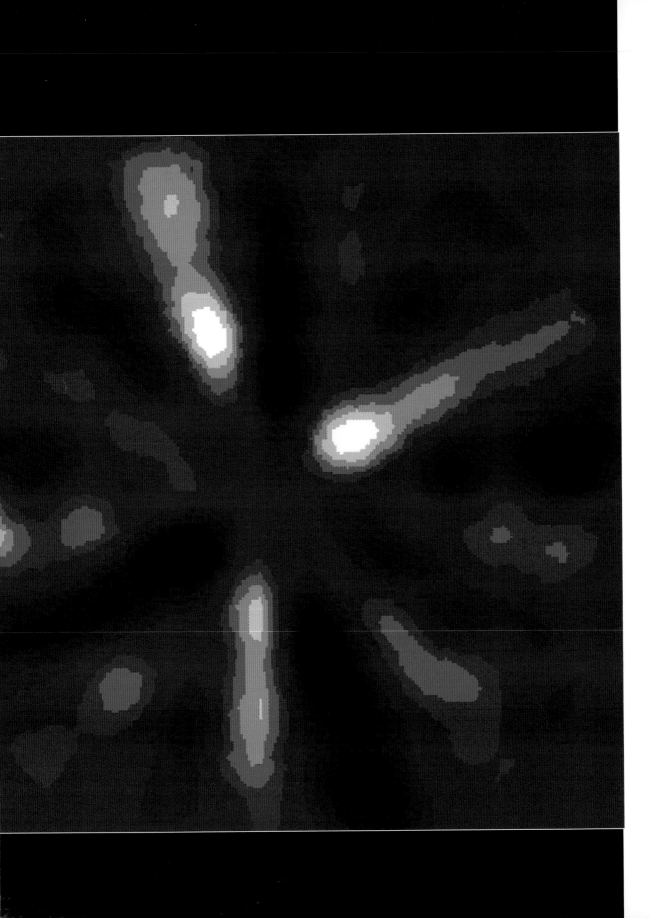

Darwin and TPF is the only way we foresee of detecting all the larger and terrestrial-sized planets around other stars, and getting a general overview of entire alien solar systems. This information would be invaluable for gaining a true picture of how planetary systems form by providing a large, and hopefully, statistically good sample of solar systems.

The next stage of the mission is where Darwin and TPF would distinguish themselves. Having located the terrestrial planets, the basic spectra of the planets can be found. At infrared wavelengths this would give the temperature of the planet (from the wavelength at which most of the light is coming) and the presence of an atmosphere. A planet without an atmosphere will just reflect the starlight towards us, whereas a planet with an atmosphere will have some of that light absorbed by carbon dioxide, thus producing the characteristic lines we see in the spectra of Venus, the Earth and Mars. This would limit the next stage of the search where Darwin and TPF would look more closely at terrestrial planets with atmospheres. This stage of the search would only take about 200 hours of observing for each planet, limiting the number of systems to about eighty. Another feature of using an interferometer is that the spectra of all the planets that the telescope is looking at can be obtained at the same time, thus shortening the search time considerably.

Another signature in the spectrum would be due to the presence of water, formed if the planet has liquid water. We assume that if a planet has an atmosphere and is at the right temperatures (between about 0 and 100 degrees Celsius), then liquid water should be present. What would be detected is water vapour in the atmosphere rather than lakes and oceans, but a significant amount of water vapour probably means that it is also present on the planet's surface.

The final, and potentially most exciting stage is the search for ozone lines. Any planet, like the Earth, with ozone in its atmosphere contains large amounts of oxygen in its atmosphere and, as far as we know, this must imply the presence of life. Darwin and TPF can also observe a line produced by methane which, on the Earth, is mainly due to herbivores breaking wind (hence it has been called the 'fart line'). Unlike ozone, which is an indication of life, methane could also be produced by volcanic activity. This last stage requires approximately 800 hours of observing

The Solar System
This simulation shows what Darwin would see of our own Solar System from 30 light years away. In this image the Sun has been cancelled out and Venus, Earth and Mars are visible as white blobs. Everything else is not real, and are just artifacts of the interfermetric observing technique.

A spectrum of the Earth
The above graph is a plotted simulation of what Darwin would be able to find out about the Earth from 30 light years away. The key ingredients are carbon dioxide (CO_2), which shows there is an atmosphere, as well as water (H_2O) and ozone (O_3), which reveals there is life.

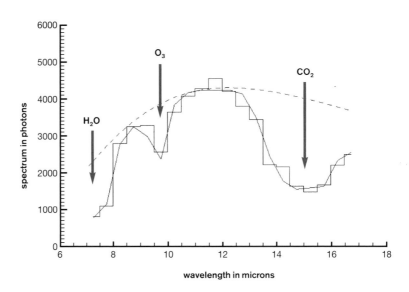

time per planet and so the number of planets that could be searched thoroughly in two years is limited to about twenty.

At the same time as exciting information is gathered on possible life-supporting terrestrial planets, Darwin and TPF can also obtain the spectra of gas giants. This would allow us to see whether they are similar to the giant planets in our Solar System. Distant and cooler gas giants are fainter at infrared wavelengths making them more difficult to see, however. Uranus and Neptune would be at the very limits of the telescope's imaging capabilities but their presence should still be detectable.

We already have an idea where planets are most likely to be located around other stars, and it is in this 'habitable zone' that Darwin and TPF will concentrate their search. Beyond Darwin and TPF is a new generation of instruments, currently being developed for the detection of new planets (see box page 137).

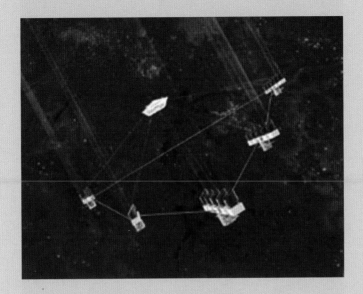

Terrestrial Planet Imager

TPI is going to be a super interferometer that will actually be able to image Earth-like planets.

A space craft like the one above would be able to produce images showing continents and oceans, cloud cover and icecaps from tens of light years away.

Beyond the Darwin and TPF missions NASA have designed the Terrestrial Planet Imager (TPI). This would be the ultimate planet-detecting mission. It will be made up of several eight-metre telescopes flying individually, at a distance of thousands of kilometres apart. They would act together as an interferometer to image other planets.

Darwin and TPF would image planets as points of light. The apparent size they have in the example image is due to their light being smeared by the telescopes and does not represent their real size. The TPI would be able to image other terrestrial planets with high resolution, identifying features like ice caps and oceans on the planets.

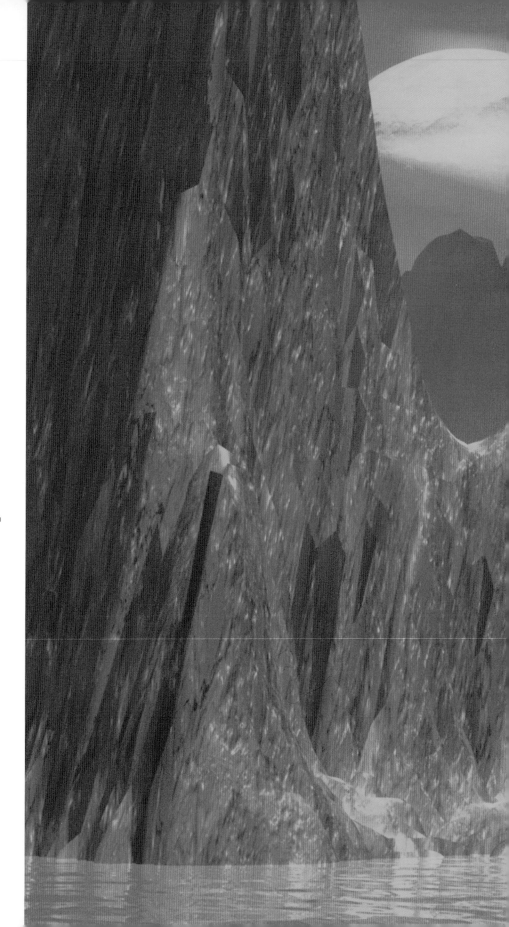

Alien vistas

An idea of the view from an extrasolar terrestrial planet. Suitable for life, the planet has an atmosphere and liquid water on the surface. Just the sort of world Darwin and TPF will hopefully find in the not too distant future.

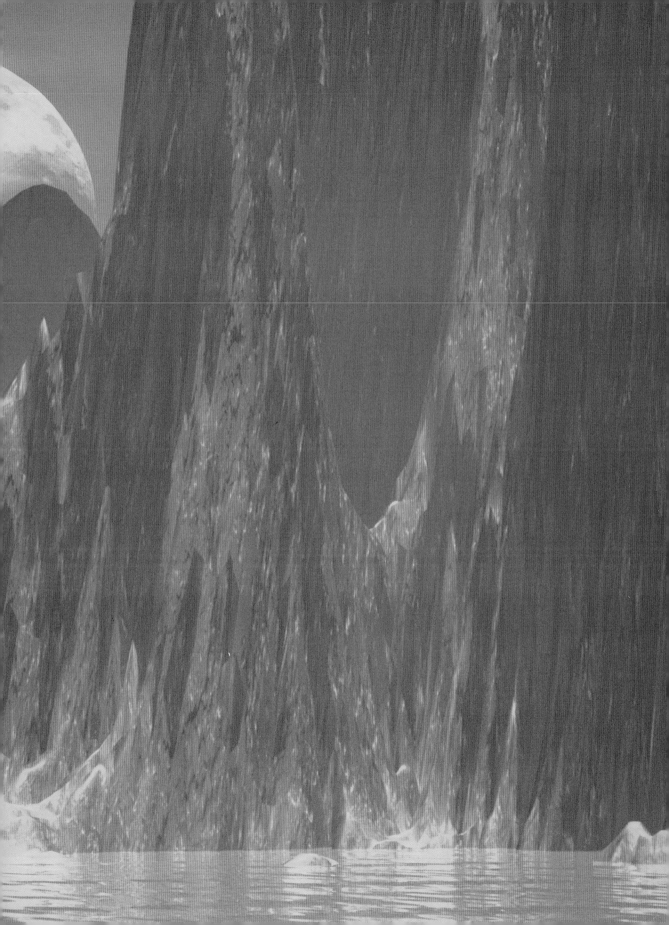

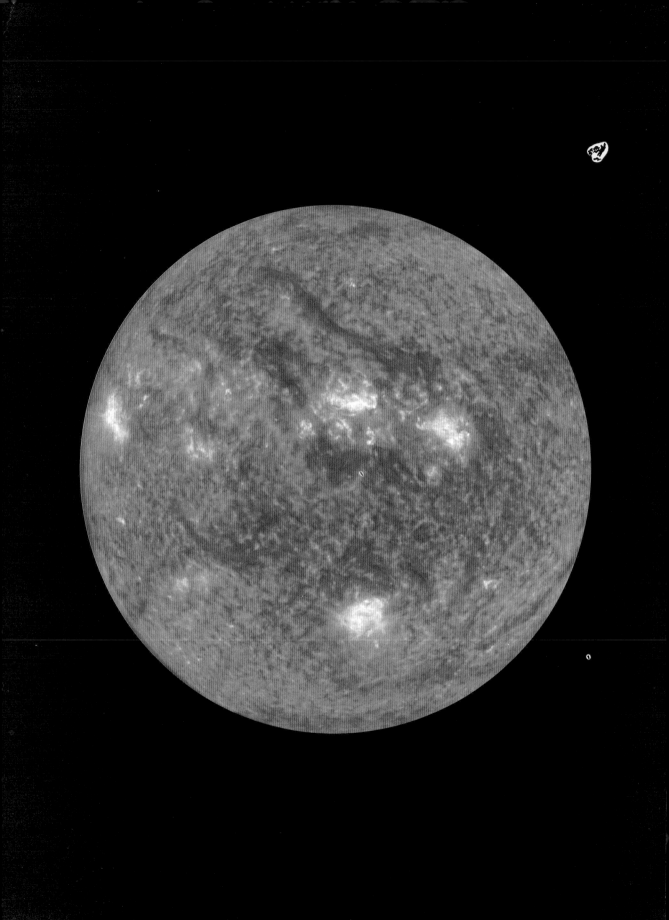

homes from home

4.1   Is anybody out there?

The question of life on other planets, and the possibility of intelligent life, is not, of course, new. For over a century science fiction has populated the Universe with all manner of weird and wonderful creatures (many of them speaking English with a suspiciously American accent). In fact the idea that the Universe is full of planets bearing intelligent life can be traced much further back to the ancient Greeks. Lucretius wrote that it is '…in the highest degree unlikely that this Earth and sky is the only one to have been created'. The ancient Chinese also suggested the same idea. The idea of a 'plurality of worlds' fell into disfavour in the west when the Earth-centred Universe of Ptolemy became the cosmological doctrine. Some, who claimed that there might be other worlds and other forms of life, notably Giordano Bruno at the end of the sixteenth century, were persecuted and killed by the church. With the Enlightenment, the idea that we may not be the only life in the Universe became popular again. It was not until the twentieth century, however, that a scientific approach was taken to this question.

Alien life form
The classic images of an alien being are human-like in appearance. We think it very unlikely that any alien life form would look anything like us.

The modern scientific investigation of the possibility of life and intelligence beyond the Earth started in 1961, when the astronomer Frank Drake (who we will meet again later) formulated what is now known as the Drake equation (see box page 145). Frank Drake originally produced the Drake equation as a means of estimating the number of civilizations in space that may be trying to communicate with other civilizations, to drive the Search for Extra Terrestrial Intelligence (SETI) projects. (These are so interesting they are discussed in more detail in section 4.3.)

The first few factors in the Drake equation are reasonably well known from astronomy. The Galaxy has roughly 300 billion stars and, as we noted earlier, the sort of stars we want for life are F-, G- and K-type stars which make up about 10 per cent of all stars. However, two-thirds of all stars are in binary systems which are probably not suitable for stable planetary orbits over billions of years. This reduces the number to ten billion stars. If we assume that life will take as long to develop to a high state everywhere else as it did on the Earth then a star must be more than four or five billion years old for multi-cellular life. This reduces the number of stars by a factor of about two, to only five billion possible stars. This is still a lot of stars.

It is at this point, however, that the Drake equation becomes rather suspect. The first three factors are known from astronomy, but the remaining factors are all currently guesswork.

We now know that approximately 20 per cent of stars locally have planets around them. These planets are all, as we have seen, massive gas giants and certainly not suitable for life as we know it. What we need to know is the fraction of stars with terrestrial planets around them. These terrestrial planets need to be in the habitable zones of their parent star to provide the conditions we think life needs to evolve (probably most importantly, liquid water).

While proponents of the idea that life is common have cited recent discoveries of other planets as proof that planets are common, these very discoveries cause problems. Stars with hot Jupiters and Jupiters in elliptical orbits are unlikely to have planets suitable for life. The reason for this is that the migrating gas giant and the terrestrial planet will interact due to gravity and will eject terrestrial planets forming in the habitable zone. The same sort of interaction would not allow terrestrial planets to have stable orbits if the system contains a Jupiter-type planet on an elliptical orbit. It may well be that the 20 per cent of stars around which we have already found planets are not the place to look for terrestrial planets in the habitable zone. This, of course, still leaves the other 80 per cent as possibilities. In only a few years, however, with the results from astrometry missions and transit detectors we will have a fairly good idea of the fraction of stars with terrestrial planets in their habitable zones, and so this factor should not provide problems for too much longer.

Estimates of the fraction of suitable planets on which life evolves range from 100 per cent to nearly 0 per cent (it cannot be exactly 0 per cent as we exist). Does life always evolve when the conditions are right, or is it an incredibly rare occurrence? Hopefully before 2020, if Darwin and/or TPF find life-bearing planets, we will have an idea if life in the Universe is common. If these missions fail to find life then the question remains open. These missions are, after all, only looking at stars within about fifty to one hundred light years – a small fraction of the Galaxy. Other life in that area would mean life is common, but no other life could just mean that life is only fairly common rather than very rare.

A larger uncertainty covers the development of civilizations. Is intelligence a survival advantage that will always evolve? Or does our technological civilization arise from an amazing set of incredible coincidences?

Using very low probabilities for the last three terms of the Drake equation can show that advanced civilizations are very rare. On the other hand, high probabilities can show that the Galaxy is packed with advanced civilizations. Can we tell which of these is most likely?

This leads us to the Fermi paradox. The famous physicist Enrico Fermi asked the question: If the Galaxy is full of advanced civilizations why are they not here already? We have no good evidence of past or present contact with an alien civilization (the claims of ufo-ologists aside), but Fermi claimed we should have met them if they existed. The key to the Fermi paradox is the amount of time it would take an advanced civilization to colonize the entire Galaxy.

The number of civilizations in our Galaxy can be estimated from

The number of civilizations in our Galaxy can be estimated from

$$N = R_* \, f_p \, n_e \, f_l \, f_i \, f_t \, L$$

where N is the number of civilizations in our Galaxy, R_* is the rate of formation of suitable planets, f_p is the fraction of these stars with planets, n_e is the number of suitable planets per planetary system, f_l is the fraction of those planets where life develops, f_i is the fraction of these planets where intelligent life forms evolve, f_t is the fraction of these planets where technology develops, and L is the 'lifetime' of communicating civilizations.

Frank Drake
The scientific search for extraterrestrial life was pioneered by the astronomer, Frank Drake.

Our current technology is not far from allowing us to travel to the stars (as we explain in section 5). Even travelling very slowly from star to star (so that each journey takes hundreds or thousands of years) we would be able to visit every star in the Galaxy in a mere few million years. Compared to a human lifetime that is a huge amount of time, but compared to the age of the Galaxy, or even the evolution of the human species, it is a very short amount of time indeed.

Imagine that we build a fleet of five spaceships to colonize five nearby planetary systems. Travelling slowly it might take 300 years to reach those stars then another 700 years might pass before the colony is able to produce more ships. Then each colony produces five more ships and sends those out to five more systems. In 1,000 years we would colonize five systems, in 2,000 years we would colonize twenty-five systems, in 3,000 years we would colonize 125 systems, and so on. At this rate we could colonize the entire Galaxy by visiting every star in one million years. Fermi's paradox was that if any civilization had developed before us, they surely would have done this. An expansion into space would allow that civilization to

Enrico Fermi
One of the most
convincing arguments
against the existence of
any technologically-
advanced extraterrestrial
civilization was provided
by Enrico Fermi.

exploit many more resources and it would survive a catastrophe on any one plane. Surely then these aliens would have visited the Solar System and found a prime piece of real estate on the third planet. Fermi reasoned that as this has not happened we are the only advanced civilization in the Galaxy.

Recently, the Fermi paradox is on firmer ground. We have seen how we will be able to detect life on planets around other stars without even visiting them. An alien civilization would surely also be able to do this, allowing them to select only interesting stars with planetary systems and life-bearing planets. It is difficult to argue with Fermi. It has been argued that advanced alien civilizations would leave any developing worlds alone (the Galactic Zoo hypothesis). This is stretching credulity a little, as even if one civilization had such high morals, we cannot expect every civilization to. We doubt that humanity would be so magnanimous about not exploiting a resource such as a planet with life.

An intriguing possibility is that we are the result of an alien visit that seeded the young planet Earth with life, and then left it to develop by itself. Why an alien intelligence would just seed the planet with one type of single-cell organism from which everything else would develop is not explained. It would make more sense to seed the planet with more advanced life forms, thus avoiding the need to wait billions of years before any interesting life form evolves. This leads us, at least, to agree with Fermi that there are no other advanced civilizations in the Galaxy. This is not the same as saying that there is no other life, but just no other life capable of colonizing the Galaxy. But why should this be so? Is there something special about the Solar System that means life, or rather, intelligent life, could evolve here preferentially?

4.2 Is the Solar System special?

The main argument of this book is that there is nothing special about our Solar System. We believe that planetary systems like ours are common in our Galaxy, and that life probably exists on very large numbers of planets orbiting other stars, although the question of intelligent life is, as we shall see, much more open. However, for the moment we will play Devil's advocate, and consider arguments which have been put forward to suggest that the Solar System really is an unusual kind of planetary system.

Comet hoover One particularly distinctive feature of our Solar System is that it contains one giant planet, which both dominates all of the other planets and lies well away from the Sun. This is distinctly different from the pattern seen in other planetary systems discovered to date (**see page 70**), although, since the search techniques that have been used would not detect systems like our own, this is not evidence that our Solar System is unique. The large planet to which we refer is, of course, Jupiter. It has a mass of 0.1 per cent the mass of the Sun, which although small is still 318 times the mass of the Earth, and more than twice the mass of the total of every other planet and moon in the Solar System. Jupiter played a large part in creating the Solar System, as we know it today. During the epoch of planet formation, as we have seen, there were very many rocky objects, similar to the asteroids we see today, which roamed the inner part of the Solar System, colliding and sticking together to form the planets. Originally, these objects moved in a variety of criss-crossing orbits, many of them highly elliptical, closely approaching the Sun and then retreating out into the Solar System before approaching the Sun again. However, the gravitational influence of Jupiter stabilized these orbits, making them more circular and ejecting many of the more unpredictable lumps of rock out of the Solar System. Without the presence of Jupiter, there would be a lot more asteroids in Earth-crossing orbits today, and more frequent collisions (perhaps every hundred thousand years or so) of the kind that brought an end to the era of the dinosaurs sixty-five million years ago. Under those circumstances, it is hard to see how intelligent life could ever have developed on Earth.

String of pearls
Comet Shoemaker-Levy 9 was ripped apart by Jupiter's gravitational pull before crashing into the planet. The cometary fragments were referred to as a string of pearls because of their linear flight pattern.

Even today, Jupiter acts as a kind of gravitational shield, collecting many of the comets that stray into the inner Solar System. This was dramatically highlighted in 1994 when the comet Shoemaker-Levy 9, broken up into a 'string of pearls' by a previous close encounter with Jupiter, impacted onto the giant planet. The images of the resulting impacts, some of which blew a temporary hole in the atmosphere of Jupiter as large as the Earth itself, were shown live on global TV and the Internet. It graphically illustrated how conditions for life in the inner Solar System might be difficult without Jupiter collecting such cosmic fragments.

The influence of Jupiter is seen most strongly in the asteroid belt between Jupiter and Mars. It was thought that an asteroid might have been the debris from a former planet that had exploded. However, today astronomers explain the asteroid belt as fragments of a planet that failed to form. In the original disc of material from which the planets formed, there was about enough material to make four rocky planets the size of the Earth between the orbits of Mars and Jupiter. As this material began to stick together into rocky bodies, they were disturbed by the gravitational pull of Jupiter. Many of the bodies were sent out into the depths of space, but many were also sent plunging towards the Sun. Computer simulations suggest that at the end of this process there may have been as many as eight super-asteroids, each as big as Mars, left over in the present-day asteroid belt. Indeed, it may well be that Mars itself is one of these objects. Most of these large objects broke up when they collided with one another because of the influence of Jupiter's gravitational pull on their orbits. However, at least one had a dramatic influence on the evolution of the inner Solar System.

A giant moon The puzzle of the origin of the Moon troubled astronomers until late in the twentieth century. Computer simulations combined with evidence of the Moon's interior structure from the Apollo missions finally solved the mystery. The Moon has a remarkably thick crust, which makes up 12 per cent of its volume, compared to the Earth, whose crust makes up less than 0.5 per cent of its volume. The Moon also does not have a core, unlike the Earth, which has a dense, iron-rich core.

Impact on Jupiter
The huge black scar that can be seen in the top left quadrant of Jupiter is larger than the Earth. It was caused by just one of the fragments of cometary Shoemaker-Levy 9. The total energy released in the impact was greater than 10,000 times as powerful as the world's entire nuclear arsenal.

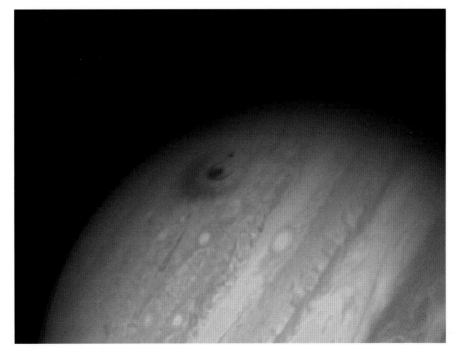

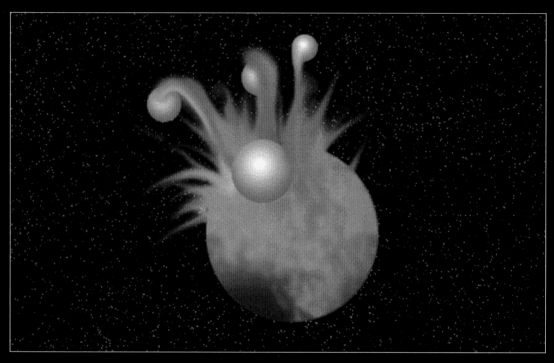

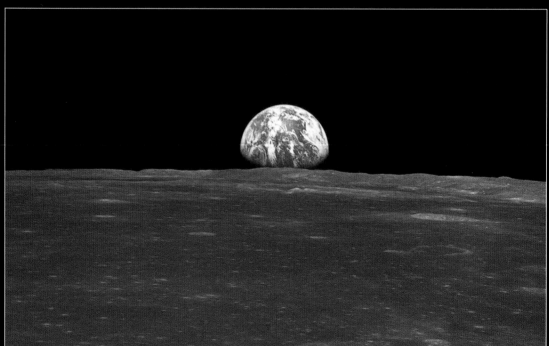

The big splash
Early in its history, the Earth collided with a Mars-sized object. The rocks thrown off from the Earth coalesced together only a few centuries after the impact to form the Moon.

The explanation now favoured to explain these differences and the origin of the Moon is that the young Earth was struck a glancing blow by an object about the size of Mars. This would have happened within half a billion years of the formation of the Earth, and in the collision both the entire crust of our planet and all of the impacting object would have melted, ejecting debris into space to form a ring around the Earth. Any metallic core that the impactor possessed would have merged with the Earth, sinking through the molten layers to enrich our own core. The ring around the Earth would be largely made up of crust material, and it was from this ring that the Moon then formed. It also explains why the Earth has such a huge Moon, with a diameter of 3,476 kilometres and a mass of 1.2 per cent the mass of the Earth. The Moon almost ranks as a planet in its own right, and the Earth-Moon system resembles a double planet, rather than a planet and its Moon (the relative size of the Moon, compared with the Earth, is ten times greater than the relative size of Jupiter, compared with the Sun).

It may be fortunate for us that such a large Moon did form. There are many desirable features of our planet that are directly associated with the presence of such a large satellite. Any dense atmosphere around the young Earth would have been removed in the impact, thus helping to insure that no runaway greenhouse effect (such as exists on Venus) ever developed here. As a direct result of the glancing impact the Earth also rotates on its axis much more rapidly than Venus does. The twenty-four hour cycle of day and night helps to even out temperatures around the globe, which seems likely to have been a beneficial thing for the evolution of life. Even the tilt of the planet, which is the cause of the cycle of the seasons, was probably produced by the impact. There is no reason to think that seasons themselves are a good thing for life (life seems to manage very well in the tropics), but the presence of the Moon has acted as a gravitational stabilizer, which prevents dramatic changes in the tilt of the Earth. In contrast, Mars, which lacks a large, stabilizing Moon, can experience dramatic wobbles which cause sudden climate changes. Such an effect would certainly not be good for advanced life on Earth. The presence of the Moon is also responsible for large tides in the oceans of the Earth.

Earth rise
The impact that led to the formation of the Moon was a lucky accident. The gravitational force of the Moon has a stabilizing effect on the Earth, which in turn has benificial effects for life on our planet.

Without the Moon there would still be tides, because of the gravitational influence of the Sun, but they would not be so big. Life first emerged in the oceans of the planet, and only later moved onto the land. It is easy to see that the presence of swampy tidal regions, which were covered by water twice a day and dried out in-between, provided a natural stepping stone between sea and land, first for plant life, later for amphibians and then other animals. Of all the seemingly special features of our Solar System which have encouraged the emergence of intelligent life on Earth, the only one which really might make us special is the presence of the Moon.

Earth's relationship with the other planets In contrast to the Moon, Mercury has a very high density, implying that it is very rich in the kind of material found in the Earth's core. In essence, the Moon is all crust and no core, whereas Mercury is all core and no crust. This can be explained naturally if Mercury underwent a major impact early in its life, this time from an impacting body hitting the planet head-on. The blast wave from the impact would have swept evenly around the planet, ejecting most of its outer layers completely, and leaving behind the dense core. The gravitational pull of Jupiter may have played a part in both processes. However, by concentrating the hazards of such impacts into the first few hundred million years of the history of the Solar System, it has also helped to insure that conditions stable enough for the evolution of intelligent life have existed on the Earth since that time.

◀ **Our massive companion**
The Moon is very large proportionally to the size of the Earth as moons go. As well as stabilizing the Earth's axis, the Moon's gravitational pull creates strong tides and makes the length of a day on Earth longer than it would otherwise be.

Is our solar system unique? Until we find a variety of different kinds of planetary system elsewhere in the Galaxy, we will not be able to say whether such conditions are common or rare. Our own view, supported by computer simulations of what happens to dusty discs around stars like the Sun, is that there must be a wide variety of planetary systems. Some of these (like the ones already detected) may have giant planets relatively close to their parent stars, but some (like our own) may have small rocky inner planets with large outer planets acting as gravitational shields. It would be just as unlikely for our Solar System to be unique as it would be for it to be the archetype of all other solar systems.

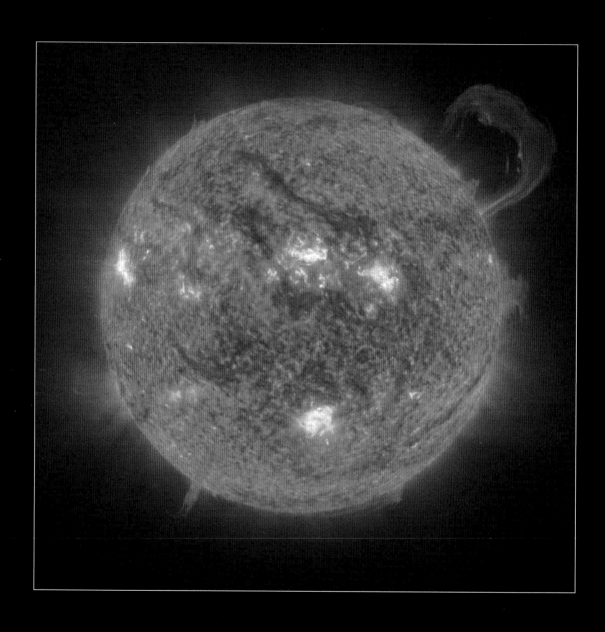

As we have mentioned earlier (page 80), most of the stars that have planets have a concentration of heavy elements (heavier than helium) as large as, or even greater than, the Sun's. This implies that stars must have a high concentration of heavy elements for planets to form at all.

This would make the Sun rather special. The amount of heavy elements in the Galaxy has steadily increased over time because planetary nebulae and supernovae produce them. The oldest stars in the Galaxy have virtually no heavy elements and most stars forming at the present time have a concentration equivalent to the Sun. This is what is so strange – the Sun is very rich in heavy elements for its age. When the Sun formed five billion years ago it was one of the most metal-rich stars in the outer Galaxy.

Could this mean that the Sun was one of the very first stars able to form planets? Life on Earth could be the first (or at least one of the first) to exist in the Galaxy. This would solve Fermi's paradox – we have not seen other civilizations because we are the first. Life may be common on terrestrial planets, but a terrestrial planet is needed to start with.

The Sun may also be unusual in that it is very stable. There has been very little variation in the energy radiated by the Sun over billions of years (after the initial heating when it was young, it has remained reasonably constant). The Sun is also relatively inactive; from Earth we see solar flares and an apparently active Sun, but from a distance of a few light years the Sun would appear to be quite inactive. Both of these characteristics of the Sun are exactly what are required for life: a stable constant heating, and no activity which would strip the ozone layer away and increase the flow of dangerous radiation onto the surface.

Many other stars that we observe are not as well behaved as the Sun. Red dwarfs flare regularly, many stars pulsate and vary their radiation output, and massive sunspots (or rather star-spots) have been seen on several other stars. All of these observations suggest that many stars are not as quiet and well behaved as our Sun, and this makes them more difficult places to develop life.

Solar flare
This image of the Sun shows a solar flare erupting from the top right-hand side of our local star. Solar flares are outbursts of very energetic charged particles that can disrupt communications and cause black-outs when they reach the Earth. Fortunately for us the Sun is relatively stable and its flares are weak compared to those of many other stars.

A stellar nursery
New stars are forming
throughout the Milky Way.
These stars are generally
rich in heavy elements
and may be better suited
to forming terrestrial
planets than older stars.

5

searching for civilizations

5.0 SETI so far

Even if we discover Earth-like planets elsewhere in the Milky Way, that will still not be proof that intelligent life exists anywhere except on Earth. Conversely, though, if we could detect direct signals from just one extraterrestrial intelligence, then we would know that conditions suitable for intelligent life exist elsewhere, without going to all the effort of building telescopes capable of seeing other Earth-like planets. The Search for ExtraTerrestrial Intelligence (SETI) has been taken seriously by astronomers since the end of the 1950s, when it became clear that existing radio telescopes had the capacity to detect signals from intelligent alien life forms, provided that such signals were directed towards Earth. As relatively little effort is needed to search for such signals, it seemed then, and still seems today, worth making that little effort in view of the potentially enormous reward that might result.

Giuseppe Cocconi and Philip Morrison first showed that if other civilizations in our Milky Way had radio telescopes and receiver systems similar to the ones that we had developed by 1959, then it would be possible for them to communicate with us. The main problem, both at the time and at the present time, is deciding which radio wavelengths are most likely to be used for communication. The first idea was to use a wavelength of twenty-one centimetres, which corresponds to the natural wavelength of radiation from hydrogen gas. Hydrogen is the most common element in the Universe; it is present in large quantities in clouds of gas between the stars, and it is of particular interest to radio astronomers who use the radio emissions from this hydrogen to map the structure of our Galaxy. It seems very likely that any radio astronomer will study the Milky Way using detectors operating at this wavelength, and it is therefore logical for any civilization attempting to contact other civilizations to send a signal at a wavelength of 21 centimetres.

Within a year of this suggestion the American astronomer Frank Drake developed the first receiver designed to search for such signals. He used a radio telescope at Green Bank in West Virginia to look for signals at a wavelength of 21 centimetres coming from two nearby stars that resemble our Sun: Tau Ceti and Epsilon Eridani. He called the modest

survey (it ran for just four weeks) Project Ozma, after the Queen in the Oz stories. The project was unsuccessful, but it was a first step towards the search for intelligent life.

The 21-centimetre wavelength is right in the middle of the waveband most suitable for this kind of study. For wavelengths longer than about 30 centimetres, the sky is full of objects emitting radio noise and it would be difficult to detect intelligent signals against the background noise. It would be like trying to photograph the stars in broad daylight. For wavelengths shorter than about one centimetre, because of effects due to quantum physics, it is very difficult to produce a signal that carries very much information. In addition, radiation from water vapour in the atmosphere of the Earth interferes with the lower wavelengths in this band. Therefore, the wavelength band favoured by all SETI searches so far is between about three and thirty centimetres, which corresponds to frequencies between ten and one Gigahertz (GHz). The 21-centimetre wavelength is equivalent to 1.4 GHz. The most modern radio receivers can tune into a band 0.1 Hz wide (the bandwidth), which means that in

signalling to the stars

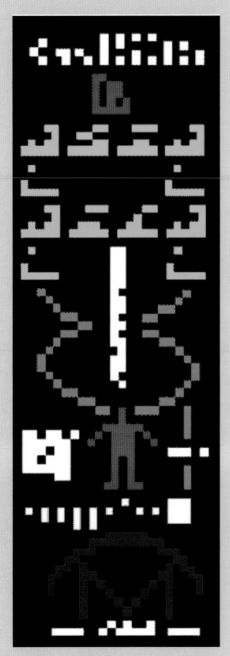

Globular Cluster M13
This cluster was the first target (albeit an odd choice) for a human attempt to communicate with other intelligences. Due to its vast distance from Earth we will have to wait 50,000 years for a reply.

The message
On the right is the message sent to M13. It contains information on DNA (the spiral in the middle); a picture of a human; and (at the bottom) a picture of the radio telescope used to send this message.

As well as listening out for signals from other intelligent life forms, we have already started to broadcast our existence to the Universe. To some extent this has been inadvertent – for example,

TV signals, have been emitted into space for more than half a century. However, much more powerful signals have been sent from the Earth into space.

The giant radio telescope at Arecibo has been used, amongst other things, as a radar transmitter and receiver to map the surfaces of Venus and Mars. This involved sending powerful radio pulses towards those planets, and detecting the reflected radio waves. Some of the radio pulses would have travelled past the target planets and into the depths of space. Any civilization with comparable radio astronomy technology to our own and which just happens to lie in the right direction might one day detect these radio pulses sent form Earth.

In 1974, the Arecibo telescope was also used to send a deliberate signal in binary code (the string of zeros and ones used by computers) towards a cluster of stars (known as M13) in the direction of the constellation Hercules. If the binary message is converted into a picture, with a white square for every zero, and a black square for every one in the message, it gives an image which any sufficiently intelligent alien could use to obtain information about the people who sent it. Even if those aliens could not decode the message, it would be obvious to them that it was a signal from an intelligent life form.

But do not hold your breath waiting for an answer – the Hercules cluster is 25,000 light years away, so there is no way we could get a response until 50,000 years after the signal was sent in the twentieth century.

The choice of M13 was rather strange. In the 1970s it was suspected that planets might require a high concentration of heavy elements to form (as we now think may well be the case), but M13 is rather poor in these elements, far less rich than the Sun. In addition M13 is a very dense cluster of stars, with the stars packed closer together than in our part of the Galaxy. This means that close encounters between stars will be far more common and these encounters would disrupt planetary systems causing planets to be ejected. All things considered, M13 is fairly unlikely to contain intelligent life forms.

this favoured window alone there are one hundred billion radio channels open to communication. Detecting a signal from aliens is not the problem, but choosing which wavelength to tune in to can be.

The fact that radio telescopes used for SETI are essentially identical to radio telescopes used in other areas of astronomy has enabled a completely different approach to the search for extraterrestrial intelligence. There is a great demand for the use of existing radio telescopes, from the many different kinds of astronomer wishing to use them, and a researcher who wants to carry out a search for extraterrestrial life forms has to take their turn in line. However, there is always the possibility of a gap in the use of the telescope. For example, when one astronomer has finished with a telescope and before the next user gets started, or when a user is switching from one of phase of an observational programme to another. Researchers at the University of California at Berkeley realized that they could make use of this otherwise 'dead' time. They have set up a programme known as Serendip in which a computer is hooked up to a radio telescope (any radio telescope whose owners are willing to play along) and waits patiently for a time when nobody else is using the antenna.

When that time occurs, the computer takes over the system and automatically records all incoming signals with a detector originally operating simultaneously on 64,000 channels. The Berkeley team have absolutely no say in where the telescope is pointing at the time, or over which range of frequencies it is tuned in to. However, this may be an advantage, because it means that they are not ignoring any particular part of the sky, or any particular part of the radio spectrum, on the grounds of sub-conscious human prejudice. The longest run that this system had was using the 100-metre Green Bank telescope in West Virginia, where serendip operated for two years before the radio dish collapsed due to old age. Fortunately nobody was hurt in the accident. Although no signal from extraterrestrial intelligence was confirmed, the system worked so well that it has led to the construction of more powerful versions operating on millions of separate channels.

The most ambitious seti programme yet attempted was carried out by NASA using an instrument known as a multi-channel spectrum analyser (MCSA) attached to the Arecibo radio telescope in Puerto Rico. This was simultaneously capable of observing just over eight million radio channels, each with a bandwidth of 1 Hz. The original intention was to use the system to search for signals from 1,000 target stars, to a distance of one hundred light years from Earth. This sounds impressive, but remember the Milky Way contains several hundred billion stars and that each star in the projected programme would only have been studied for half a minute at any particular wavelength during a ten-year survey. However, even this modest step was never taken. Congress cancelled funding for the NASA project in the early 1990s, after only 200 hours of work, when just 24 stars had been investigated. But this is far from the end for SETI, and several new projects are likely to get underway during the next few years.

5.1 The future of SETI

The ambitious NASA programme to monitor 1,000 nearby stars for signals from extraterrestrial intelligence was cancelled shortly after it began operating. However, the special receiver built for the project survived the cuts and may be reborn as part of an appropriately titled project: Phoenix. This project is part of the work of the privately funded SETI Institute based in Mountain View, California. The President of the Institute is Frank Drake, who carried out the first search for extraterrestrial intelligence more than forty years ago. The MCSA, upgraded to cover 14 million channels, could be installed at the Parkes radio telescope in Australia to scan the southern skies, and then moved back into the Northern Hemisphere (probably to Arecibo, or to Nancay, in France) to look at the northern skies.

Similar, but more modest, surveys are also being planned or carried out by astronomers such as Paul Horowitz at Harvard University .
However, there is an alternative to this kind of brute force survey of the entire range of the radio spectrum from one to ten GHz. Just as the pioneers identified the 21-centimetre radio emissions from hydrogen as a likely wavelength on which intelligent beings might broadcast, so their successors have identified other 'magic' frequencies.

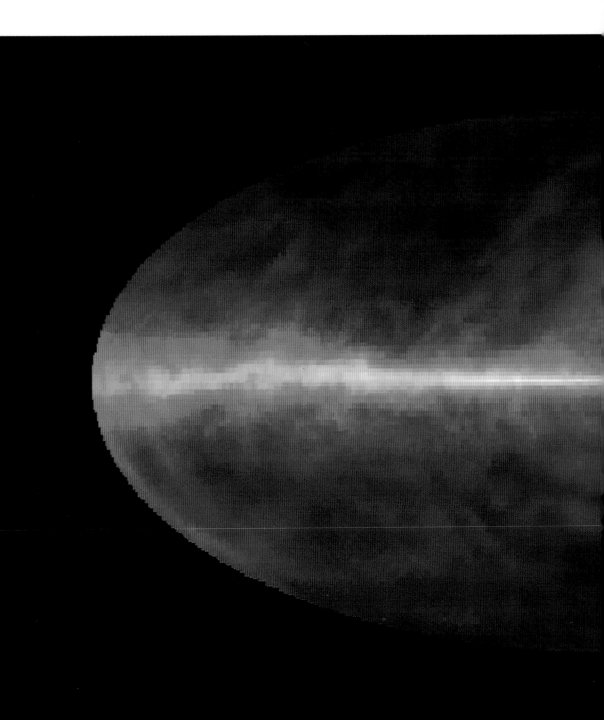

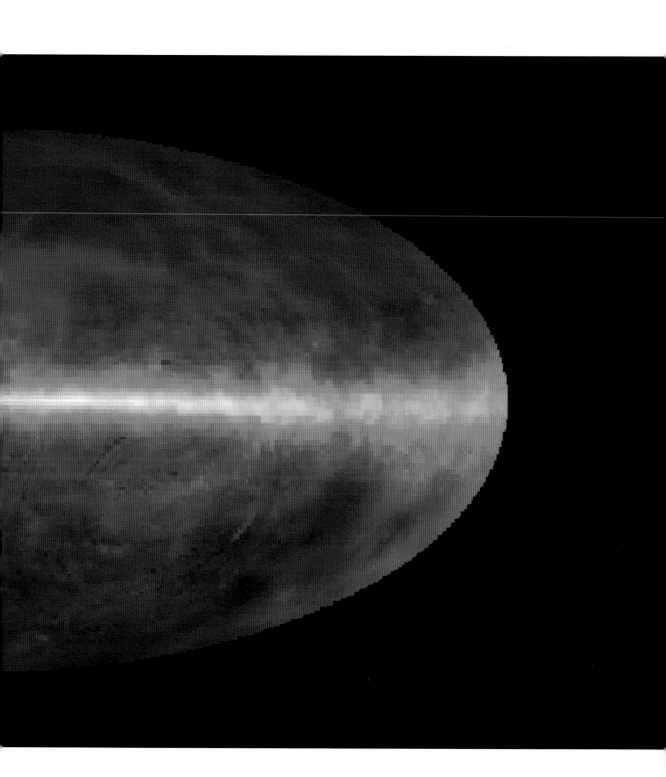

The Water Hole The radiation from hydrogen at 21 centimetres is called a 'line' by analogy with the lines produced by different elements in the spectrum of visible light (see page 37). The second strongest radio line detected from the clouds of gas and dust that lie between the stars of the Milky Way occurs around 18 centimetres. In fact, there are four distinct lines close to this frequency, all produced by radiation from a compound consisting of single hydrogen atoms linked together with oxygen atoms, forming what is known as a hydroxyl radical. The hydroxyl radical is like a molecule of water, which has had one hydrogen atom removed. Because water is such an important requirement of life, this suggests to some radio astronomers that any life forms like ourselves will broadcast at wavelengths between that of hydroxyl and the hydrogen line – i.e. in a range from 18 to 21 centimetres. This has been called 'the Water Hole', as an analogy to the way different species of animals congregate at water holes on Earth.

Looking for signals broadcast deliberately by aliens is not the only way to search for extraterrestrial intelligence. As we have mentioned, the signals we have broadcast inadvertently are already spreading into space and fill a circle of 50 light years in radius. Other alien civilizations are also likely to be broadcasting in this way, whether they have any intention of making contact with us or not. The great advantage of searching for such signals is that, if other intelligent life forms do exist, there are likely to be many such signals to detect. The signals are also likely to be broadcast all the time, even if the level of information that they contain is no higher than that of daytime television on Earth. The great disadvantage of searching for such signals is that because they are not beamed through space in one direction (like the beams from the Arecibo radio telescope) they will be very weak by the time they are detected on the Earth. In order to eavesdrop on such inadvertent interstellar transmissions, we will need larger and more sensitive radio telescopes and detectors than current technology has constructed. This does not mean, however, that we could not construct such detectors if we wanted to. Back in the 1970s, Bernard Oliver, of the company Hewlett Packard, carried out a study for NASA of the kind of equipment that would be required. Although never intended for construction at that time, it was given the name Project

The sky seen in radio waves
This image shows the sky in 21-cm radio waves. The bright band across the centre is the Milky Way. At this frequency the sky is relatively free from bright objects and other celestial objects, like obscuring clouds, that make finding extrasolar planets such a difficult pursuit.

Cyclops, because it would be like a large radio eye looking out into space. Radio astronomers have long since developed the technique of linking many different antennae so that they mimic the properties of one large radio dish. The scale of project envisaged by Oliver would have involved at least 1,500 separate antennae, each consisting of a dish with a diameter of one hundred metres, and receiving equipment and computers necessary to make them act like a single antenna.

One of the attractions of such a project is that the project could start before all 1,500 antennae had been built. A single radio dish would be necessary for the project to begin, with the others being hooked up to the system once they were completed. Clearly, such a project would be very expensive by the standards of scientific research. However, to put the cost into perspective, the entire system (during the period that Project Cyclops was being carried out) would cost about the same as the United States spent in just three months on the war in Vietnam.

In case you are wondering whether we really do have the technology to carry out such a project, the techniques involved are already in use at a radio observatory known as the Very Large Array (VLA), in New Mexico. This array is made up of 27 antennae, each of them with a diameter of 25 metres, constructed along a Y-shaped railway track in which each of the three arms is 15 kilometres long.

The track is bigger than it needs to be relative to the size of the antennae in order to allow each of the individual elements in the array to be moved so that a range of different spacing between the antennae can be achieved. This is a useful tool in radio astronomy. The Cyclops array, consisting of 1,500 antennae, could be packed together as closely as possible, and would cover a circular region of desert with a radius of less than five kilometres (and an area of roughly 80 square kilometres).

The Allen Telescope One instrument now being planned will involve several hundred small antennae packed into an area of about 10,000 square metres. The project is being established with a donation of US$12.5 million (about half the total cost of the project) from two Microsoft millionaires, Paul Allen (one of the founders of Microsoft) and Nathan Myhrvold (former Microsoft chief technology officer). It is (rather optimistically) intended to become operational in 2005.

The array, known as the Allen Telescope, is another brainchild of the SETI Institute, and is being built in northern California. It will be able to detect the signals from a dozen star systems at a time, and, like all these instruments, can also be used for conventional radio astronomy work. If you are interested, you can have one of the radio dishes named after you for a donation of US$50,000.

The United Nations announced the latest variation on this theme in the summer of 2000. An agreement in principle to build a radio telescope called the Square Kilometre Array (SKA) was signed in August 2000, and at the present time the project is due for completion, probably in Australia, in 2015. The smaller Allen Telescope will be useful as a test bed for the techniques involved in the SKA, as well as carrying out its own search for extraterrestrial life forms. Between them, the two telescopes will cover both hemispheres of the sky. Several countries (Australia, Canada, China, India, the Netherlands and the US have signed up so far) will share the projected cost of US$1,200 million.

As many of the examples we have discussed highlight, one of the major problems with any search for extraterrestrial intelligent life forms involving radio telescopes is the amount of computer power that is required to analyze the flood of data received from the antennae. Or rather, it is not so much the power of the computer, but the large amount of computer time that is required.

The analysis required to detect an intelligent signal actually requires nothing more complex than a home computer. At the end of the 1990s astronomers realized that the Internet now provided a means for literally millions of home computers to get involved in SETI. They created the SETI@home project, which enables anybody who wants to connect their computer via the Internet to a system collecting data from a sky survey being carried out by the Arecibo Telescope. In the spirit of the Serendip projects, the computer system makes use of any time available when the telescope is not actually engaged in its regular astronomical work. It will gather data at whatever frequency the telescope is operating on at the time, and from whichever part of the sky it happens to be pointing at that time. Anyone who subscribes to the project downloads a screen saver, which includes a small programme dedicated to analyzing a tiny fraction of the information obtained from Arecibo.

The software carries out this task while the home computer is not in use (provided that it is switched on!) and is programmed to report any intelligent signal that it discovers. Well over one million home computers now spend at least part of each day looking for signs of intelligent life in the Universe, and it is quite possible that if extraterrestrial life forms are detected it will happen on a computer sitting on the desk of some computer nerd. You can get the screen saver from: www.setiathome.ssl.berkeley.edu.

SETI screen saver
Download the SETI screensaver and your computer can be used to analyse radio signals searching for artificial communications above the noise.

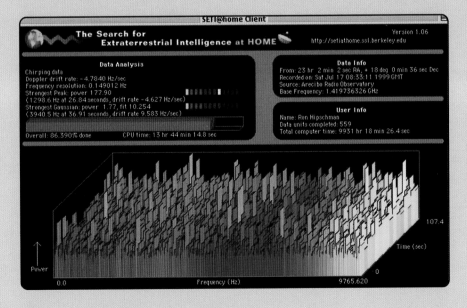

The SKA will be the most sensitive astronomical instrument yet built on Earth with a collecting area of one square kilometre (one million square metres). However, the SKA is not solely dedicated to the search for extraterrestrial intelligent life forms. It will also be used to study the cosmic background radiation (a faint hiss of radio noise left over from the Big Bang), to study the way stars and planetary systems form, and for other astronomical work. It is not quite on the scale of Project Cyclops, but will still involve several hundred separate antennae.

The biggest criticism of systems such as Project Cyclops and the SKA is perhaps provided by the work of Paul Horowitz of Harvard University. This suggests that the special part of the search for extraterrestrial intelligence is not the telescope, but the computer system that it is connected to. With the rapid progress in computer technology and miniaturization in recent decades, Horowitz developed a system so compact that it could be taken from place to place and connected to any telescope that was available. In its original form, the system operated on eight million channels, each with a bandwidth of 0.05 Hz. The aim of the Harvard team was to explore the range of frequencies centred on 1.4 GHz (21 centimetres in wavelength). By using a spread of frequencies around this value they could detect signals that were broadcast precisely at this wavelength, but have been shifted slightly by the Doppler effect (see page 55). This shift is caused by the motion of the planet the signals are broadcast from, and by the motion of the Earth through space. The project was originally funded by a private organization called the Planetary Society, and like everybody else who gets the chance, in the 1980s Horowitz took his system to Arecibo for a trial run. The trial worked well, and he began considering what to do next. At the same time an old radio telescope with a 26-metre diameter dish belonging to Harvard University and the Smithsonian Institute became obsolete and was closed down by the radio astronomers. With the aid of the Planetary Society, the Harvard team was able to take over the telescope and adapt their system into a permanent project, known as Project Sentinel. In the second half of the 1980s the situation improved still further when Steven Spielberg donated US$100,000 to upgrading the project into a system known as the Mega channel Extra Terrestrial Array (META), operating on eight million channels. Each day, the telescope is left pointing in the same direction, while the Earth rotates on its axis, so it sweeps around a strip of the sky. The next day, another strip is scanned, and so on. All of the sky visible from Massachusetts can be covered in 200 days, and then, like the painting of the Forth Bridge, it is time to go back to the beginning and start again. META relies on the very low probability that a signal will be received from the right part of the sky, at the right frequency, when the telescope is pointing in the right direction. However, the longer we keep on searching the better the prospect of finding any such signals.

5.2   Getting there ourselves

The hard way to find planets orbiting other stars is to physically go to the stars and look for them. The distances between the stars are so huge, however, that it is hard to see how we can explore the Galaxy ourselves in the way that people like Christopher Columbus explored the Earth. Even the nearest stars to the Sun are several light years away, which literally means that light takes years to travel between them. The speed of light is an absolute limit on the speed with which any object can travel through space, and the speed of light is enormous – 300,000 kilometres per second. Science fiction fans will know that if you could travel at a significant fraction of the speed of light then, for you, time would slow down, so that even a journey between two stars one hundred light years apart could in principle be covered within a single lifetime. Back home, many thousands of years would have passed, as graphically portrayed in the movie Planet of the Apes, but there might well be people intrepid enough to undertake such a one-way voyage into the future. Unfortunately, in order to achieve the speeds required, the input of energy would be absolutely prohibitive. As we shall see, it is the limit of present-day technology to accelerate very small probes, incapable of carrying astronauts, to quite small fractions of the speed of light.

However, it is possible, even with existing technology, to send unmanned probes to the stars. It just takes a very long time. If we are prepared to wait to reap the benefits, then in terms of the lifetime of a

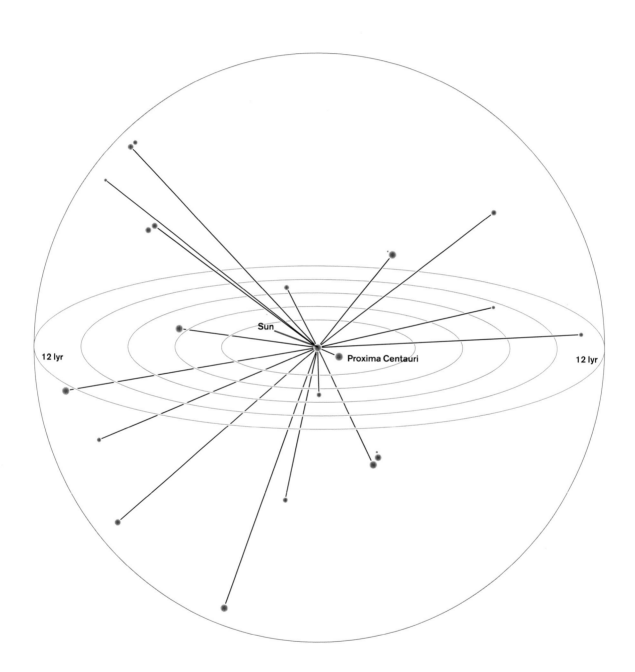

Sun

Proxima Centauri

12 lyr

12 lyr

human civilization, or of a country like the United States, it might well be worth the effort. However, the effort is only going to be worthwhile if we are reasonably sure that we are sending our probes towards stars that do possess planets. Therefore this phase of the exploration of the Galaxy follows on from everything we have discussed so far.

There are at least three technologically feasible ways of sending probes to the nearest stars. The first, although feasible, is so extravagant that it borders on the insane, and is highly unlikely ever to be put into practice. Back in the 1950s the same scientists who invented the hydrogen bomb proposed a design for a spacecraft which could be propelled by ejecting a series of nuclear devices from its rear (rather like laying eggs) and exploding them one after another. The debris from the explosion would strike a strong 'pusher plate' at the back of the spacecraft and propel it forward. The physicist Freeman Dyson has calculated that a spaceship with a total mass of 400,000 tonnes could carry a payload of 20,000 tonnes (supporting hundreds of crew members) to Alpha Centauri in about 130 years. Even this extravagant vehicle, dubbed Orion, would reach a maximum velocity of only one-thirtieth the speed of light, and would still take longer than an average lifetime to reach its destination. We do not think that anything like this will ever be built.

However, the second type of deep space propulsion system has already been tested and shown to work. It uses the opposite approach to that used in the nuclear spaceship, but instead of brute force and nuclear explosions it involves a weak, but continuous rocket pushing the spacecraft forward. Using electricity to push charged atoms (ions) of heavy atoms such as mercury or xenon backwards, the ion drive gently pushes the spacecraft in the opposite direction. The amount of thrust generated is about the same as the pressure exerted on your outstretched hand by a sheet of paper laid flat on your palm. Such a weak thrust can build up the very high speeds required because the engines can run for a very long time. Eventually, it will deliver ten times as much thrust for every kilogram of fuel it carries as conventional chemical rockets. The system has been tested on the NASA probe Deep Space 1, which was launched in October 1998. By the autumn of 2000, it had been running its engine for more than 200 days in space.

The only system that has operated longer is an identical ground-based version of the engine run in a test chamber at the Jet Propulsion Laboratory, which was scheduled to complete 15,000 hours of continuous running by the end of 2000. Deep Space 1 was intended to run for a comparable time, until the autumn of 2001. The engine uses just one hundred grams of xenon per day, and less than half one kilogram every four days. By the end of its mission to rendezvous with Comet Borrely, the little engine will have changed the speed of the spacecraft by more than 11,000 kilometres per hour.

Deep Space 1 could never be a true interstellar probe, because it gets the electricity it needs to run its ion engine from sunlight. However, it would be quite straightforward to develop a similar engine, which obtained its energy from a nuclear reactor; the so-called nuclear electric propulsion (NEP). The immediate restriction on developing such space probes is that people are understandably reluctant to allow nuclear reactors to be launched from Earth, in case of accidents.

However, it would be possible with present-day technology to build such a system using a 150 kilowatt nuclear reactor (only about as powerful as an electric fire) which could run the ion engines for several years. Although genuine interstellar travel is still some decades off (at least), there is a proposal on the drawing board for an interstellar precursor mission (IPM). This could combine a mission to Pluto with a probe that would continue out into space, at least 1,000 astronomical units from the Sun, to test the feasibility of such engines. The whole system would have a mass of 32,000 kilograms, of which the IPM probe would be about 1,500 kilograms and the Pluto orbiter probe, on the back of the IPM, would have a mass of about 500 kilograms. The remainder of the mass would include a propulsion system, at least one nuclear reactor, and reaction mass for conventional rockets used to get the spacecraft on its way.

The design for a genuine interstellar probe that we like best, however, would not have to carry any of this extra weight. Many scientists have drawn up plans for probes that would 'sail' between the stars with the aid of intensely powerful laser beams directed at the vehicle from the Solar System. The power source for the spacecraft would remain in the Solar System and only the minimum weight would have to be carried to

Planet of the Apes? Travelling through space to other planets will not be complete science fiction in the near future. We think it unlikely that we will ever find the Planet of the Apes however.

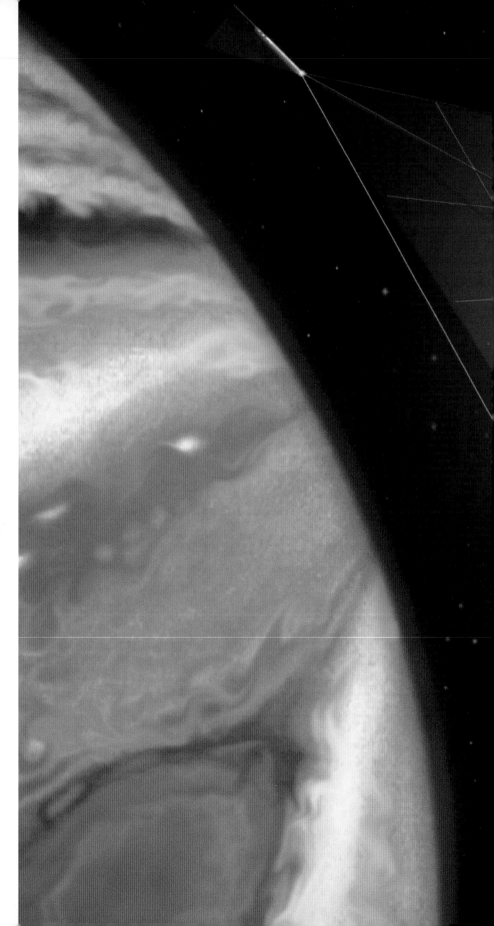

Solar sail

A solar sail such as this would be a cheap way of travelling around the Solar System. With the use of giant lasers to push it forward, a solar sail could travel between the stars.

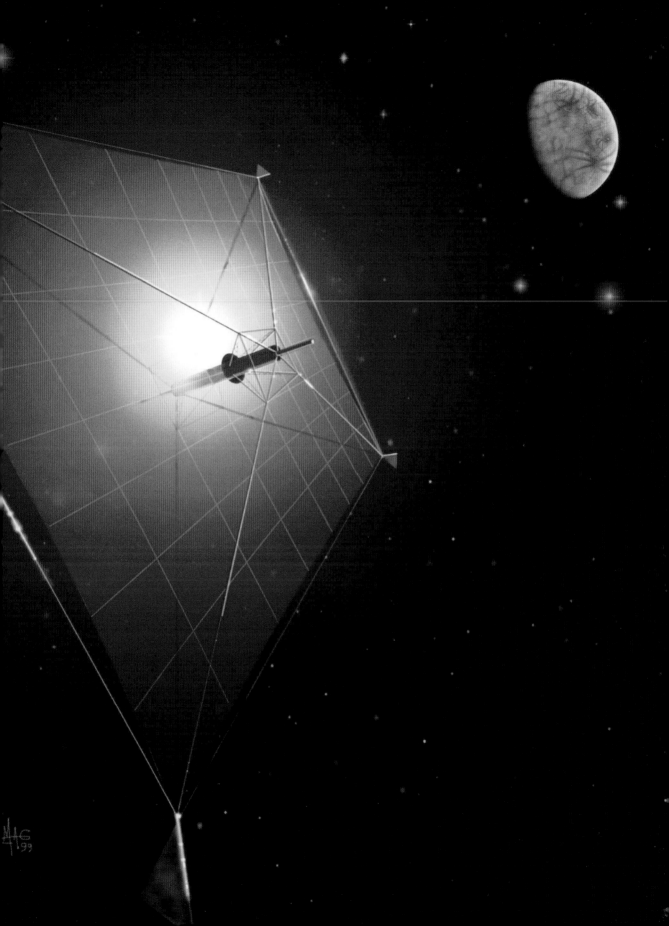

One of the most extraordinary possibilities deriving from modern technology is the idea that only one or a few interstellar probes might be necessary in order to explore the entire Galaxy by remote control. The mathematician John von Neumann proved half a century ago that it is possible in principle to construct a machine, which has the capability of constructing an exact replica of itself. This idea now seems obvious in the light of the development of computers. However, this idea has been developed further into the so-called 'Santa Claus machine', which can build anything.

Such a machine could be set down on the surface of the Moon, or on a barren planet such as Mercury, and could use material from the rocks and the energy from the Sun to make anything it was programmed to make. The software required is very easy to transport, it does not even have to be carried by the machine, but could be beamed by laser from Earth once the machine was in place. If the machine were small, it would have to be programmed to build the tools, to build the factory and to build the other equipment, required to build a replica of itself. But this would only involve more time, more solar energy, and more rock.

Whichever approach to interstellar travel you favour, all that would be required would be to send one of these von Neumann probes, as they are usually called, into another planetary system, where it could replicate itself and send each replica to explore other stars and planets. For example, with the star sailing approach, the huge investment required to develop the powerful lasers to send the probe on its way need not be so huge after all. All that has to be built on Earth is a Santa Claus machine, which is capable of sitting on the surface of Mercury and doing all the hard work. It would then be straightforward for the system to give one or more probes the same instructions to fly to nearby stars, where they would seek out other barren planets, or use material from asteroids to set up new bases and repeat the whole process. While they were doing so, of course, they would be sending back information to Earth about their discoveries.

This approach to exploring the Milky Way does require a little patience, but not a lot more than was required by our recent ancestors when the Earth itself was being explored a few hundred years ago. For a star sailing probe, the time required to travel from one star to the next is a few decades, one century at most. If the technique is to work at all, it would surely take no more than another century to build the systems required to send not just one but several more probes onto further stars.

Even being pessimistic, the generation time required to move on from star to star is only a couple of hundred years. Our descendants or ourselves would start to get information back from the first star visited within a few years of the arrival of the first probe, because signals would travel back to us at the speed of light. If by then we had discovered Earth-like planets orbiting nearby stars, the von Neumann probe could be instructed to send its next generation of probes to look at them.

With time, for the entire future of human civilization, more and more information about more star systems would be sent back. The most important piece of information that could ever be sent back in this way, would be the news that there are other intelligent life forms in the Universe and that we are not alone. This would surely justify both the cost of the project and the patience required seeing it through.

John von Neumann ▶
The mathematican, John von Neumann showed that it would be possible to create a self-replicating machine. Such machines could be used for long-distance space travel.

another planetary system. The best site for the lasers would be in orbit around the planet Mercury. The gravity of Mercury would act as a tether, stopping the lasers from drifting into space, and they would get their energy from the Sun. A lot of sunlight would be required. In order to accelerate a large space probe at the equivalent of one-third the force of gravity at the surface of the Earth, the laser would have to carry 43,000 terawatts of power. (One terawatt is one million million watts.) This is some forty times the total amount of power generated today in all the power stations on Earth. It would enable the light sail to reach half the speed of light in just over one-and-a-half years, after which the laser could be switched off and the space probe left to coast the rest of the way. It would reach the star Epsilon Eridani in just 20 years.

The sail required to make use of all this laser power would be enormous – the more optimistic projections stipulate sails of about 1,000 kilometres in diameter. In order to slow down on reaching the target star, the inner portion of the sail, 330 kilometres across and carrying the instrument pod, would be detached and lag behind the remaining larger portion. Provided the laser beam in our Solar System had been switched on again at the right time, its light it would then reflect off the larger portion of the sail. This would make it go even faster, and would reflect back onto the front of the inner portion of the sail, thus slowing it down. The whole process would undoubtedly work, but it would require an enormous input of energy and a lot of engineering to get a single probe on its way to the stars. This may be worthwhile, however, if no greater effort is required to make a whole series of probes than to make one (see box page 180). Such ideas may all seem like science fiction, and it is surely no coincidence that one of the leading proponents of star sailing, Robert Forward, is also a leading science fiction writer. But it is worth remembering that Leonardo da Vinci had the vision to imagine flying machines back in the fifteenth century and models built to his designs have shown that if he had had access to modern technology (in particular, the internal combustion engine), he could have made them work. Interstellar space probes may seem no more remarkable to our descendants in a couple of century's time than aircraft do today.

If we do discover other terrestrial planets with atmospheres, or even life, it is very unlikely that these planets will be immediately habitable for ourselves. Even if oxygen is present in an atmosphere the chances are that we could not breathe it. We are very sensitive to the amount of oxygen in the air – too much or too little and we will die. If it were possible to go back in time to the Jurassic age of the dinosaurs, we would need breathing apparatus as the Earth's atmosphere contained insufficient oxygen for our lungs.

There is a speculative, but genuine, branch of science concerned with terraforming or planetary engineering. This is the process by which we can artificially make a currently uninhabitable planet habitable (or, at least, less uninhabitable). Most of the work on terraforming so far has been aimed at making Mars a better home for Earth-based life. Even if we are never able to make the Martian atmosphere breathable, thickening it would increase the surface temperature and pressure making it possible to walk about in less than full space suits.

The idea for terraforming Mars is to try to melt the ice caps (actually frozen carbon dioxide caps). The frozen carbon dioxide would enter the atmosphere and, as it is such a good greenhouse gas, it would trap heat, thus melting more of the ice caps in a continuous process. Possible methods to do this are massive orbiting mirrors to focus more sunlight onto the ice caps or covering the ice caps with a dark dusty layer which would increase the amount of heat they absorb (ice is very good at reflecting sunlight away). A lot of serious thought has gone into investigating how Mars could be terraformed.

The problem is reversed with Venus as the atmosphere is too thick and it is too hot. It has been suggested that aiming comets at Venus would blast away some of the atmosphere. Given enough comets a significant fraction of the atmosphere could be removed. This method would also have the added advantage of introducing water to the cooler planet.

The process of terraforming would not have to make a planet an exact duplicate of Earth (that may hardly ever be possible). It would be good enough to make a planet more suitable for us. If the air pressure is high and the temperature is reasonable then plants and animals may be able to live unprotected. In addition any failure in our life support systems would be less likely to be a catastrophe.

These ideas, which could well be used in the future in our own Solar System could be extended to extrasolar planets when they are colonized. This would make the Galaxy a far more hospitable place.

A terraformed Mars
With a thicker atmosphere, Mars could become wet again. In this artist's impression, we can see the huge northern ocean that would form, were this to happen.

making ourselves at home

The Ultimate goal
Eventually we would like
to think that the human
race could go in person
to explore the extrasolar
planetary systems that
we are now finding.
Here people are shown
exploring a planet close
to the Orion Nebula. In
space, spectacular views
are a certainty.

Useful Web addresses

The discovery of new planets occurs on an almost weekly basis and we are sure that our list of planets will not be complete by the time you are reading this book. The best place to look at the latest news of planets is the internet which has some great resources. Some of our favourite sites are:

The Extrasolar Planets Encyclopaedia
http://cfa-www.harvard.edu/planets/
A great site updated often with news of new planets and a very up-to-date list of all the planets found so far. Also has a great links page to other extrasolar planets sites.

Catalog of Extrasolar Planets
http://www.astronautica.com/
A graphics heavy site with an excellent set of images of various aspects of extrasolar planets. For the amateur astronomer it has a guide to finding the stars around which planets have been found. Some of these are visible with the naked eye for those without telescopes (51 Peg is pretty easy to find on a dark night). Excellent links page as well.

Astrobiology Web
http://www.astrobiology.com/
A site dedecated to the new scientific study of possible life in space. Rather specialised but dense with information. Good links off to other astrobiology pages.

Extrasolar Planet Search
http://exoplanets.org/
Geoff Marcy's and Paul Butler's (planet finders extrodinaires) site. An up-to-date list of new planets. Their almanac of planets shows the radial velocity curves used to find many of the planets.

Extrasolar Visions
http://www.jtwinc.com/Extrasolar/
A lovely site by John Whatmough where he provides artists impressions (based on good science) of what extrasolar planets may look like, and the views for their surfaces.

Darwin & TPF
http://ast.star.rl.ac.uk/darwin/
http://tpf.jpl.nasa.gov/
The homepages of the Darwin and TPF missions. Generally quite specialised but some information for the public is available on both sites.

SETI
http://www.seti.org/
The home of SETI, a good site with lots of general information.

seti@home
http://setiathome.ssl.berkeley.edu/
The place to get your seti@home screensaver which will set your PC to work while you are not using it analysing data looking for SETI radio signals.

Index

Figures in italics refer to pictures.

Acknowledgments

THE PUBLISHERS would like to thank both Simon Goodwin and John Gribbin for all their expertise and pictorial contributions; NASA for the invaluable educational resources thet offer in all fields of astronomy and space exploration; Judith Menes for all her hard work in preparing the index of this book; Mike Shayler for his proof-reading skills; and Caroline Thomas for finding many of the pictures that are listed below.

PICTURE CREDITS: cover: Jeff Hester and Paul Scowen (Arizona State University) and NASA; p.2 NASA; pp. 6-7: Tony & Diane Hallas/Science Photo Library; p.10: Image produced by F. Hasler, M. Jentoft-Nilsen, H. Pierce, K. Palaniappan, and M. Manyin. NASA Goddard Lab for Atmospheres - Data from National Oceanic and Atmospheric Administration (NOAA); p.14: NASA/JPL/Northwestern University; p.16: NASA/JPL; p.17: NASA/Ames Resesarch Center, Image No. G72-0411; pp.18-19: Image produced by F. Hasler, M. Jentoft-Nilsen, H. Pierce, K. Palaniappan, and M. Manyin. NASA Goddard Lab for Atmospheres - Data from National Oceanic and Atmospheric Administration (NOAA); p.20: NASA/JPL/ Malin Space Science Systems; P.21: NASA/JPL: p.22: NASA/JPL; pp. 24-25: NASA/JPL; pp. 26-27: NASA/JPL; p.28: " CORBIS; pp.34-35: NASA, The Hubble Heritage Team, STScI, AURA; p.37 (top): Nigel Sharp (NOAO), FTS, NSO, KPNO, AURA, NSF; pp.38-39: pp.40-41: R. Sahai and J. Trauger (JPL), the WFPC2 Science Team and NASA; p.45: (left) A. Dupree (CfA) and NASA, (right) C. Barbieri (Univ. of Padua), and NASA/ESA; p.47: (all photos) Margarita Karovska (Harvard-Smithsonian Center for Astrophysics) and NASA; p.51: (bottom)NASA/STScI; p.52: NASA/SAO/CXC; pp.56-57: Jeff Hester and Paul Scowen (Arizona State University) and NASA; p.60: R. Sahai and J. Trauger (JPL), the WFPC2 Science Team and NASA; p.61: NASA/ESA/STScI; p.62: Jonathan Blair/CORBIS; p.65: (bottom) Prof. Michel Mayor, Observatoire de Genève; pp.68-69; John Foster/ Science Photo Library; p.74: C.R. O'Dell and S.K.Wong (Rice University), NASA; p.78: C. Burrows and J. Krist (ST ScI) and NASA; p.79: (both photos), M.J. McCaughrean (MPIA), C.R. O'Dell (Rice University), NASA; p.82: Susan Terebey (Extrasolar Research Corp.), and NASA; pp.84-85: James Gitlin/STScI AVL; p.86: NASA/JPL; p.89: W.Couch (University of New South Wales), R. Ellis (Cambridge University), and NASA; pp. 94-95: SIM/JPL/NASA; p.103: © Raymond Gehman/ CORBIS; p.104: NASA/JSC; p.110: NASA/JPL; p.109: All Rights Reserved Beagle 2; p.111: Dr Ken Macdonald/Science Photo Library; pp.112-113: NASA/JPL; p.114: S. G. Gibbard et al. (IGPP, LLNL); p.118: © Paul A. Souders/CORBIS; p.119: (top) © Joseph Sohm: Chromo Sohm Inc./CORBIS, (bottom) © Temp Sport/ CORBIS; p.120: © Jeffrey L. Rotman/CORBIS; p.128: Copyright Alcatel; p.126: H. A. Weaver (Applied Research Corp.), HST Comet Hyakutake Observing Team, and NASA; p.130: Rutherford Appleton Laboratory; p.132: GSFC/NASA; pp. 138-139: Roger Harris/Science Photo Library; p.142: John Foster/Science Photo Library; p.148: NASA/JPL; p.152: NASA/JPL; p.149: H. Hammel, MITand NASA; P.150: (top) NASA, (bottom) The Electronic Universe Project, University of Oregon; p.154: NASA/JPL; p.158: NASA/JPL/CALTECH; p.162: Yuugi Kitahara/ NASA; p.163: Frank Drake (UCSC) et al., Arecibo Observatory (Cornell, NAIC); pp.166-167: J. Dickey (UMn), F. Lockman (NRAO), SkyView; p.168: Ronald Grant Archive; p.171: SETI@home, University of California, Berkeley; p.176: © Bettman/CORBIS; pp.178-179: Mark Garlick/Science Photot Library; p. 180: © Bettman/CORBIS; p.182: NASA/JPL/CALTECH; pp.183-184: Chris Butler/Science Photo Library; p. 186: © Bettman/CORBIS,p.192:NASA/JPL.